T0172421

BEHAVIOURAL
PRODUCTION

Autonomous manufacturing and cyber-physical systems are key enabling technologies of the Fourth Industrial Revolution (IR4) which are currently being incorporated into the building design and construction industries. These emerging IR4 technologies have the potential to effectively improve construction affordability and productivity, address current and future building demand, and reduce the environmental impact of the built environment. However, design approaches that make use of IR4 technologies are still relatively unexplored. While automation, such as mass production, promotes standardised design solutions, design thinking that embraces varying degrees of autonomy can lead to unique and considered approaches to design on an industrial scale.

Behavioural Production: Semi-Autonomous Approaches to Architectural Design, Robotic Fabrication and Collective Robotic Construction explores design operating through the orchestration of spatiotemporal events. A multi-agent behaviour-based approach to computation is employed in architectural design and extended to individual and swarm-based robotic methods for additive manufacturing. *Behavioural Production* seeks to expand our capacity to engage with the world at large through varying degrees of autonomy. In an industrialised world where traditional craftsmanship has been marginalised and cannot scale to meet societal needs, this book speculates a means to bring scalable forms of creativity into the act of making. This is explored through the use of materials, generative algorithms, computer vision, machine learning, and robot systems as active agents in design conception and realisation. The book presents a collection of ideas, projects, and methods developed in the author's design practices and research labs in the fields of architecture and computer science. This body of work demonstrates that engaging with semi-autonomous processes does not diminish authorship, but rather expands it into new forms of design agency that seamlessly integrate with emerging manufacturing and construction technologies whilst authoring distinctive design character.

Robert Stuart-Smith is Program Director of the Master of Science in Design: Robotics and Autonomous Systems degree (MSD-RAS) at the University of Pennsylvania, an Affiliate Faculty member in Penn Engineering's GRASP Lab and Assistant Professor of Architecture. Stuart-Smith also directs the Autonomous Manufacturing Research Lab in Penn's Department of Architecture and University College London's Department of Computer Science. Stuart-Smith's research explores generative approaches to design, robotic fabrication and collective robotic construction. Recent projects include the £1.2 million EPSRC project "Applied Off-site and On-site Collective Multi-Robot Autonomous Building Manufacturing" and a £2.4 million EPSRC research project into "Aerial Additive Building Manufacturing", involving collaborations with industry leaders Cemex, Skanska, Mace, Burohappold, Arup, MTC, Ultimaker, and others. Stuart-Smith was also a co-founder of the experimental design practice Kokkugia and now directs architectural practice Robert Stuart-Smith Design, which focuses on projects closely aligned with his research activities. Stuart-Smith's work has been published in journals including Nature, Science Robotics, AD Architectural Design, and Architecture D'Aujourd'hui, while he has given lectures and presentations at institutions including ETHZ, AA, UCL, CCA, MIT, RMIT, ICD Stuttgart, Angewandte, Strelka Institute, Paris-Malaquais, Tsinghua University, and others.

BEHAVIOURAL PRODUCTION:
Semi-Autonomous Approaches to Architectural Design, Robotic Fabrication and Collective Robotic Construction

Robert Stuart-Smith

Routledge
Taylor & Francis Group

LONDON AND NEW YORK

Cover image:
Additively manufactured ceramic column exterior skin ©Robert Stuart-Smith, Autonomous Manufacturing Lab, University of Pennsylvania. Photograph by Renhu (Franklin) Wu.

First published 2024
by Routledge
4 Park Square, Milton Park, Abingdon, Oxon OX14 4RN

and by Routledge
605 Third Avenue, New York, NY 10158

Routledge is an imprint of the Taylor & Francis Group, an informa business

© 2024 Robert Stuart-Smith

The right of Robert Stuart-Smith to be identified as author of this work has been asserted in accordance with sections 77 and 78 of the Copyright, Designs and Patents Act 1988.

All rights reserved. No part of this book may be reprinted or reproduced or utilised in any form or by any electronic, mechanical, or other means, now known or hereafter invented, including photocopying and recording, or in any information storage or retrieval system, without permission in writing from the publishers.

Trademark notice: Product or corporate names may be trademarks or registered trademarks, and are used only for identification and explanation without intent to infringe.

British Library Cataloguing-in-Publication Data
A catalogue record for this book is available from the British Library

Library of Congress Cataloging-in-Publication Data
A catalog record has been requested for this book

ISBN: 9780367463410 (hbk)
ISBN: 9780367463427 (pbk)
ISBN: 9781003028291 (ebk)

DOI: 10.4324/9781003028291

Typeset in Neue Haas Grotesk Display Pro
by Adam Blood

Graphic Design: Adam Blood, Patrick Danahy, Robert Stuart-Smith.

MIX
Paper | Supporting
responsible forestry
FSC® C013056
www.fsc.org

Printed and bound in Great Britain by
TJ Books Limited, Padstow, Cornwall

Contents

Acknowledgements

Behavioural Production primarily presents design and research work from my design practices Robert Stuart-Smith Design and Kokkugia, and the Autonomous Manufacturing Lab research teams I direct in the University of Pennsylvania's Department of Architecture and University College London's Department of Computer Science. Parts of this work were supported by:

Supporting Institutions:
University of Pennsylvania, Weitzman School of Design's Department of Architecture
University College London, Department of Computer Science
RMIT University, Architecture and Urban Design
Architectural Association School of Architecture, Design Research Laboratory

Funding Support:
Work included in this book was supported by the UK's Engineering and Physical Sciences Research Council (EPSRC) for projects EP/N018494/1 and EP/S031464/1; University of Pennsylvania's University Research Foundation, Weitzman School of Design's ARI Robotics Lab and Faculty Development Fund; RMIT University's Research Stipend Scholarship and Australian Government Research Training Program Scholarship; Cemex Global Research and Development.

Any of the opinions, recommendations or conclusions presented in this book are those of the author and do not necessarily reflect the views or policies of any supporting institution or funding body.

There are also several people I would like to acknowledge:

Much of the early work presented in this book was developed in the experimental design practice Kokkugia, which I founded together with lifelong friends Roland Snooks and Jonathan Podborsek in 2004, operating from studios in Melbourne, London and New York until 2012. Kokkugia sought to design a complex architecture, developing generative algorithmic design methodologies to realise designs that could not be easily drawn or modelled. Many of the ideas, methods and projects from this early work are especially interconnected with Roland Snooks from our years of collaboration. Thank you Roland, for your limitless drive and excitement for design, friendship and life. And Jono, thank you for bringing an edgy and stimulating approach to practice, and for adding a healthy dash of humour to every moment. Kokkugia's custom design methodologies and interest in complex systems were challenging to realise. Most of our built work required strategic implementation of traditional means of construction due to the nascent nature of digital fabrication technologies two decades ago. Since this time, access to more advanced fabrication technologies has increased, allowing my work to further the possibilities of a complex architecture through my practice Robert Stuart-Smith Design, and in two closely aligned research labs.

Thanks to support from University of Pennsylvania's Department of Architecture and University College London's Department of Computer Science, I founded the Autonomous Manufacturing Lab in both universities to pursue more time and resource-intensive research into design, robotic fabrication and collective robotic construction (AML-UCL in 2016; AML-Penn in 2017). The AML-UCL lab is co-directed by Dr. Vijay M. Pawar, who, along with a colleague in aerial robotics at Imperial College London and Empa, Prof. Mirko Kovac, are both dear friends and have been integral to some of the most pioneering work in this book. Our different fields of knowledge and experience have converged into a rich and complementary collaboration. Especially in our Aerial Additive Manufacturing research project, for which I am grateful for the scientific contributions, extensive brainstorming and camaraderie we developed together with research collaborators: Stefan Leutenegger, Chris Williams, Paul Shepherd, Richard

Ball, and each of our research teams. Thank you also to the talented and dedicated team of architects, researchers and students that were integral to the development of each project and are credited in each chapter's project credits.

The book also includes work from several years of collaboration with Cemex Global R&D's Vice President Davide Zampini, Alexandre Guarini, Carlos Enrique Terrado, Valentina Rizzo, Matthew Meyers and others. Thank you for your generous support and collaboration on research and teaching. Your interest in synthesising material innovations with new concepts for fabrication and design is inspirational and tightly aligned with the ideas and work documented in this book.

The University of Pennsylvania Weitzman School of Design's leadership and the Department of Architecture's faculty and staff practically supported this work in several ways but also encouraged my curiosity and scholarship during the last seven years. Some of the recent research could not have been achieved without the Weitzman School's fundraising and support for a new robotics lab facility. Special thanks to Dean Fritz Steiner, Chair of Architecture Winka Dubbeldam, Acting Chair of Architecture Andrew Saunders, Leslie Hurtig, Chris Cataldo, Karl Wellman, and Jeff Snyder. The Architectural Association's School of Architecture Design Research Laboratory also supported some of the early Aerial AM work described in Chapter Nine, and was the forum for my academic work for more than a decade beforehand. Brett Steele, Theodore Spyropoulos, Patrick Schumacher, Shajay Bhooshan, Michael Weinstock, David Greene, Marta Malé-Alemany, Alicia Andrasek, Tyson Hosmer, and Mel Sfeir were particularly insightful and encouraging throughout that period of work.

This book was also written during, and in part is tied to my PhD studies at RMIT University. PhD supervisors Paul Minifie and Vivian Mitsogianni have greatly impacted my perspective on this body of work and how I plan to evolve it. A special thank you to Adam Blood and Patrick Danahy for collaborating on the layout and graphics development of this book, and to Fran Ford, Hannah Studd, Alanna Donaldson, and the rest of the Routledge Taylor and Francis staff for years of dialogue and support throughout the peer review, write-up and copyediting of this manuscript. Thank you also to the many colleagues who have supported, informed and critiqued this work in formal and informal ways over several years.

Stella Dourtme, my wife and business partner, has supported me immeasurably in the writing of this book. Not only providing insightful guidance sharpened from her years of knowledge and experience working on some of the most state-of-the-art design and construction projects ever built, but also for countless small acts that ensured I could conclude this book during this lifetime – thank you! And finally, Peter and Trish Stuart-Smith provided a lifetime of invaluable personal and professional development support, while my young daughter, Niovi, has been a constant reminder of the inherent creativity and autonomy within all of us right from the start of our lives!

Preface

This book explores a behavioural approach to design that commenced more than two decades ago with the development of generative algorithmic design approaches and subsequently evolved to incorporate a similar attitude towards materials, robot systems, and their sensing and tooling capabilities. For the avid reader seeking to understand this book's place within the field of architecture, it is differentiated in its development of behaviour-based approaches to design that engage with diverse media and robot systems and in its aim of producing a distinct design character whilst addressing practical and logistical considerations of design and production.

Most uniquely, pioneering work in Collective Robotic Construction and Aerial Additive Manufacturing is presented that conceptually offers a scalable approach to design and production. This contribution is demonstrated in the author's role as the sole architect to collaborate with world-leading roboticists to write a review paper on Collective Robotic Construction in *Science Robotics,* and in co-initiating and co-leading research that produced the world's first Aerial AM – untethered aerial additive manufacturing in flight, published as a cover article in *Nature* last year. The author's contributions to these activities involved the development of a novel approach to swarm-based multi-robot mission and path planning that coordinates the distributed, concurrent actions of an autonomous collective of robots in live demonstrations and construction simulations, and the integration of this software framework within generative design workflows in order to design through the act of construction. This multidisciplinary research was undertaken through the author's leadership in both architectural and computer science fields of research, as a director of research labs in both fields[1] for the last eight years, directing government-funded research projects[2]. The work provides opportunities for situated and embodied forms of architectural design that have the potential to be more sensitive to site, community and climate, and other spatiotemporal concerns. By unifying design and construction, architecture becomes a four-dimensional concern, open to fleeting temporality or enduring continuous construction possibilities.

Additional contributions are provided in the development of design methodologies in the areas of multi-genus topological design, the integration of visual character analysis and reinforcement learning within generative design processes, the use of computer vision and machine learning for site-adaptive design, and behaviour-based adaptive manufacturing processes that engage with material via real-time robot programming and sensor-feedback from articulated robot arms and mobile robot systems. These methods pit aesthetic concerns against the environmental and economic impacts of design, while unifying design and production. The advocated unity of design and production as a shared creative act undercuts present approaches to design that necessitate value-engineering geometric rationalisation exercises and suggests a more holistic and collaborative approach to the realisation of architecture could be achieved.

The work covered in this book operates mostly within computational methods and their relation to either 3D graphical media or individual robot and swarm-based approaches to additive manufacturing. This does not mean to imply that designers must code algorithms (AI software is becoming increasingly effective at doing this), rather, the book is focused on a conceptual engagement with degrees of autonomy that the author hopes might outlive present-day workflows or technologies. Despite this, the author developed the majority of the software code for this work in several programming languages on Linux, Windows and Mac operating systems, and where employees contributed, directed them with detailed pseudocode. While this might locate the book within a "computational design" sub-field of architecture that typically seeks to produce an architecture driven by performance or geometric optimisation concerns, this would not be a perfect fit. Although the work addresses such design criteria, it was also motivated by a desire to produce highly specific design character through the development of bespoke design methodologies, and advocates for negotiated design solutions that might be sub-optimal in one or another consideration, but well-rounded to several design concerns.

While this work operates largely through bottom-up algorithmic approaches to engage with diverse aspects of our world, there is nothing puritanical in the approach. Semi-autonomous approaches are developed and leveraged to extend design agency, not to contain it. For this reason, the author shamelessly engages in intuitive, explicit 3D modelling and other top-down design activities when deemed beneficial. It is hoped this book encourages a similar open-minded approach to design in the reader.

This is first and foremost a book about design, that advocates for an expansion of considerations, methods/approaches, engagements and effects that can be achieved through design by embracing degrees of autonomy. *Behavioural Production: Semi-Autonomous Approaches to Architectural Design, Robotic Fabrication and Collective Robotic Construction* explores approaches to design that operate through the orchestration of spatiotemporal events within 3D modelling environments, materials, manufacturing processes, and individual and collective robotic systems. A multi-agent behavioural approach to design, manufacturing and construction is presented, that is systemically open and able to be employed in diverse generative processes while authoring distinctive behavioural character. *Behavioural Production* might be viewed as part monograph and partly a reflection on how one can engage in design within a world increasingly dependent on autonomous systems.

This form of material engagement, while authored, de-centres the human by extending design agency to partially operate through deliberative or causal acts by other virtual or physical entities. To better understand this condition, the work is also discussed in relation to Vital Materialism, Object-Orientated Ontology, Posthumanism, and other theoretical constructs. As society embraces the Fourth Industrial Revolution's autonomous manufacturing and cyberphysical systems technologies it will become increasingly easy to fall back on pre-prepared derivative approaches to design bundled into software subscriptions that do not encourage the designer to interrogate the premise of any particular methodology. It is hoped that this book inspires further exploration into novel design methodologies, as this is critical to ensuring a more eclectic and exciting future for design. Embracing degrees of autonomy within design enables us to go beyond automation to foster a more diverse and considered built environment.

Notes:
1. As director of the Autonomous Manufacturing Lab in the University of Pennsylvania's Department of Architecture (AML-Penn) and University College London's Department of Computer Science (AML-UCL).

2. Key projects include the £2.4m + £500k Industry Support funded Aerial Additive Building Manufacturing project (EP/N018494/1), and the Applied Off-site and On-site Collective Multi-Robot Autonomous Building Manufacturing project with £1.2m funding + £600k Industry Support (EP/S031464/1)

1 | Autonomy
Orchestrating and Partnering with External Phenomena

Behavioural Production

If you were to overhear two people in conversation mentioning the words *"behaviour"* or *"autonomous"* it is doubtful that you would imagine they were engaged in a discussion about design, manufacturing, or construction. The Merriam-Webster dictionary defines *"behaviour"* as the *"action or response of an individual, group or species"* to their environment or other forms of stimulation (Merriam-Webster 2023). For architecture to be explored through the crafting of behaviours, software, machines, materials, or other things, must essentially be tasked with participating in the conception or realisation of a design. It is in this context that the word *"production"* is used in the book's title as a broad term to encompass the creative and diverse grouping of design, manufacturing, and construction activities described in this book.

Advocating for a behavioural approach to design may at first seem strange; however, it is one of the oldest and most utilised approaches to building we are aware of. Social insects such as termites, design through the act of building, using "stigmergic" principals where each termite's building behaviour is adapted to the environment and the previous building actions of other termites (Grassé 1959; Dorigo, Bonabeau, and Theraulaz 2000). This simple bottom-up process gives rise to self-organised collective behaviour that robustly operates in diverse and dynamically changing environments to produce emergent building designs (Bonabeau et al. 2000). In contrast to such adaptive and scalable approaches, today's design, building and manufacturing industries rely on automated routines and infrastructure that cannot scale to meet present-day or future demand. Perhaps our estimated 7 million years of development (Wong and Hendry 2023) hasn't been enough to fully grasp the benefits of behavioural approaches to design, manufacturing and construction compared to the 135 million years termite species have had to develop and refine their methods (Krishna and Grimaldi 2003). Now, at the outset of the Fourth Industrial Revolution, we are starting to engage in autonomous forms of manufacturing (see Chapter Two). For our species though, these developments are not biomimetic but foregrounded by a history of computation, robotics and autonomous systems developments that stretch back at least two thousand years.

Autonomy

Automatons

In Homer's *Iliad* — one of the most legendary works of fiction — there is a story about an architect of incredible mythological powers. Hephaestus, the god of blacksmiths and fire, and son of Zeus and Hera, was the sole designer for the Olympian's mansions, each built of an indestructible bronze architecture (Butler 1970, 76; Cartwright 2019). Hephaestus also crafted numerous other tools, weapons and innovative mechanisms including self-moving automatons – maids of gold that could learn and converse, and a giant bronze automaton that guarded the island of Crete from invaders until it was killed by Medea (Henderson 2019, 195). To help forge his metallic designs, Hephaestus also built twenty self-operating and self-adjusting bellows (Henderson 2019, 197). Thus, this architect of mythology was also an innovator of autonomous manufacturing and humanoid automatons.

Homer's *Iliad* offers the first known use of the word *automaton* to describe something *'acting of one's own will'* (Henderson 2019, 189). Automatons typically employ a precise interplay of mechanisms such as pulleys, levers, and gears to produce a desired action automatically. Outside of mythology, the development of Ancient Greek architecture arose within a cultural milieu that included parallel learning and innovation in construction methods and automata. One Ancient Greek engineer living around the first century CE designed many. Documented in his book *"Pneumatics"* (Alexandria and Woodcroft 2015), Heron of Alexandria's inventions helped with the construction of buildings and in automating aspects of their use. Heron developed construction crane designs, automatically opening doors, and a house alarm, along with a human automaton called the *'Servant of Philon'* that could pour a mixture of wine and water into a guest's glass (Alexandria and Woodcroft 2015; Kotsanas 2018).

In a two-book treatise called Automaton-Making (Peri Automatopoiētikēs) (Grillo 2019), Heron describes a historical mobile automaton used for a mobile theatre, that could autonomously wheel itself into position, open its doors and act out a story with mobile figurines and scenes, before closing and moving out of sight (Grillo 2019, 3). He also outlines a design for a more capable version that could be programmed with different sets of motion sequences (Grillo 2019, 7) through a sophisticated arrangement of threads wound around a series of axles that defined durations of motion in different directions as well as pauses (Sharkey 2007). Ibn Ismail Ibn al-Razzaz Al-Jazari's 1206 *Book of Knowledge of Ingenious Mechanical Devices* describes a similar type of automaton called the *"Drinking Boat"* (alJazari and Hill 1975). Comprised of small musician figurines that could play different musical sequences if supplied with a different program, the Drinking Boat's music was defined by a specific arrangement of pegs distributed around a cylinder. As the cylinder rotated, the pegs struck levers attached to the musicians' mechanised bodies to create different motion and music effects (Sharkey 2007).

There is very little conceptual difference between these historical automata and early programmable manufacturing devices such as Joseph Marie Jacquard's *'Jacquard Loom'* (1804) or calculating machines like Charles Babbage's *'Difference Engine'* (1822) (Pask and Curran 1982, 12). Both the Jacquard Loom and Difference Engine used punch-card programs, not dissimilar to early electro-mechanical computers including Konrad Zuse's *'Z1'* (1936), and IBM's *'ASCC Mark1'* (1944) (Pask and Curran 1982, 12). This genealogy for computer science is made more explicit in John von Neuman's development of an algorithm he called a *"cellular automaton"*, that created an array of cells that operated as a finite-state machine, where each cell could be considered an agent, able to determine its own state in relation to the states of its neighbouring cells. Applied in the *Universal Replicator* project, von Neuman demonstrated that cellular automata could be employed as a self-replicating computer program (Schwartz, von Neumann, and Burks 1967). Stephen Wolfram extended cellular automata algorithmic research, producing emergent complex patterns that arise through the programming of even simpler automaton's event-based rules (Wolfram 2002).

Locally perceived autonomous acts drive more elaborate agent-based behaviours in Craig Reynolds' *"Boids"* algorithm to simulate the dynamic motion of flocks of birds or schools of fish, producing emergent collective behaviour (Reynolds 1987). Reynolds' computational approach was inspired by Braitenberg's robot vehicles (Braitenberg 1986) which can be viewed as a conceptual extrapolation of William Grey Walter's *Machina Speculatrix* (or *'Tortoise'*) robots. Walter's robots leveraged sensor-driven reactive behaviours to interact with each other and their environment – an impressive feat for pioneering 1950s robotics research (Walter 1953). Taken together, this partial sample of automata, manufacturing equipment, computers, algorithms and robots, all share a historical lineage that leverages degrees of autonomy that range from simple automated routines to fully autonomous behaviour. This book explores various means by which such degrees of autonomous behaviour become operative within both the conception and realisation of architecture, extending creativity into acts of production.

Autonomous Systems and Degrees of Autonomy
Automated and autonomous systems are not the same. In general terms, something is automated if it performs acts involuntarily based on a pre-determined set of commands, while something is autonomous if it is able to self-govern or operate independently ("Autonomous" 2022). In the context of robotics and computer science, these differences are more nuanced.

Autonomy in Robotics and Computer Science
Karel Capek first used the term *'robot'* in his 1921 theatrical production *Rossum's Universal Robots*, deriving the term from the Czech word for forced labour – *'Robota'* (Jordan 2016, 30). While Capek's biological robots were programmed to have no autonomy, roboticists such as Maja Matarić, consider a robot to be an autonomous system that can sense and act (behave) within the environment (Matarić, Koenig, and Arkin 2007, 1–2). For robotics researchers, an autonomous system can adapt its behaviour to diverse circumstances by being programmed to make different decisions in relation to perceived events. There are, however, different degrees of autonomy. In the context of mobile robot platforms (such as driverless cars or robot vacuum cleaners), degrees of autonomy often relate to how a robot perceives, navigates, determines or executes tasks within an environment. Robots are considered either fully autonomous or semi-autonomous depending on how dependent they are on external knowledge to determine and execute actions.

From the perspective of computer science, Sheridan and Verplank devised a framework comprising ten levels ranging from automation to autonomy. At level 1, a human does all the thinking and set-up before handing over execution to a computer while at level 10, a computer determines whether a task should be done, chooses and executes the task, and decides whether a human should be informed of what was done (Sheridan and Verplank 1978, 8–19). Level 10 would also include anything that approaches general-purpose artificial intelligence (AI). Throughout this book, the term "intelligence" is avoided where possible, as it all too easily implies more than what is being discussed. Although some parts of this book utilise task-specific Deep Learning approaches, outside of such cognitive or learning methods scientists in developing autonomous robot systems also use the word intelligence to distinguish more situational knowledge than automation affords. However, *from a designer's perspective it could be argued that there is nothing necessarily 'intelligent' about autonomous systems. Approaches presented in this book are simply more adaptive or creative than automated systems.*

Having established that automated and autonomous systems are not the same, what value is there in developing autonomous robot systems? There are many tasks that are too remote, microscopic, or out of reach for people to perform. There are also activities we find challenging or dangerous that require heavy lifting, dexterity, speed, precision, high-frequency repetition, visual or physical sorting, data analysis, or continuous operation. Where possible these tasks are automated; however, there are many occasions where automation is insufficient to ensure robust, repeatable, highly controlled, or variable operation. Examples include: unsupervised mobile robots avoiding obstacles as they move through an unknown or dynamically changing environment, or a manufacturing robot programmed to maintain a constant sanding pressure by dynamically responding to force-torque sensor data feedback. In these situations, for humans to extend their level of agency, they must enable software or robots to operate with degrees of autonomy. *While this may seem counter-intuitive, providing software or a robot with some degree of autonomy does not diminish human agency. Instead, human control is able to extend into dynamically determined events that would otherwise be beyond our capacity.*

Advocating for Degrees of Autonomy in Design
In consideration of the benefits of autonomous systems in the field of robotics, why develop autonomous methods for architectural design and construction? To understand needs and opportunities, Chapter Two discusses the instrumental role autonomous systems have in the emerging Fourth Industrial Revolution (IR4). The chapter also evaluates to what degree automated and autonomous approaches have already entered the building construction sector, and the level of interest, research and development taking place in today's architectural academic institutions. After establishing the societal and architectural context for autonomous construction, the book presents a conceptual approach in which design can integrate autonomous methods for practical and creative benefit.

Some architects might question whether relying on software or robotics systems for any part of the design process undermines humanitarian concerns, cultural values, or the historical and disciplinary thinking of architecture. As such, it is important to clarify that autonomous systems are not separate or detached from human cultural production; they have emerged from it. Nor will developers or users of autonomous systems be absolved of ethical or professional accountability. As we enter the era of the Fourth Industrial Revolution (IR4), our dependence on such systems is expected to grow (discussed in the next chapter). In this context, why wouldn't architecture *critically* engage with such a significant cultural and technological societal shift? After all, it is the architect's responsibility to ensure the continual evolution of humanitarian, cultural, and disciplinary concerns within architectural practice. Historically, both mass-production and mass-customisation manufacturing techniques have been bound by rigid industrial and economic standardisation practices. This standardisation arguably encouraged generalised, prosaic design solutions that constrained architectural tectonics, part-to-whole relationships, material organisation, aesthetic expression and other aspects of design. However, with the emergence of IR4 and autonomous manufacturing, these limitations are being challenged as more specific and individualised design responses become possible. In addition to the societal benefits discussed in the next chapter, engaging with degrees of autonomy allows for an expansion in the agency and aesthetics of architectural design. It also offers potential material and economic efficiencies, as well as environmental benefits.

Behavioural Production: Semi-Autonomous Approaches to Architectural Design, Robotic Fabrication and Collective Robotic Construction explores a conceptual approach to design that aims to develop agency within various elements that influence or form architecture. This includes acts of fabrication or construction. This proposition involves letting go of some direct control in order to allow designs to gain additional capacity to engage more directly with the world at large, and within a more extensive set of multidisciplinary collaborations. In this book, design operates through material agencies, where people, materials, objects, things, software code, tools, equipment, and processes contribute to design cognition (see Chapter Three). This argument wouldn't resonate though if it could not be demonstrated at least to some degree. The projects presented in this book attempt to forge a path forward, but should be viewed as exploratory, partial insights into what could be. While there is some degree of technical demonstration, there remain challenges to overcome if one is to implement this research in practice.

This book is foremost about design, and promotes an expansion of considerations, methods, engagements and effects that can be achieved through design. *Behavioural Production* explores varying degrees of autonomy within material, computational and robotic processes as an integral aspect of a creative design process. While it emphasises the inseparability of design from material and technical aspects, the book does not argue for complete autonomy or the exclusion of people from design or construction activities. Instead, an inclusive approach to people, machines, computers, materials, objects and things is advocated, where with curiosity and humility we can explore creative possibilities by engaging cooperatively with diverse aspects of our world. With this in mind, a few topics need to be addressed and discounted.

What this Book isn't About

Are We Nearing the End of Architectural Practice?
The recent emergence of large-language model-based AI chatbots such as ChatGPT (OpenAI 2023) bring software much closer to Sheridan and Verplank's level 10 – full autonomy. Together with generative AI software such as Midjourney, Stable Diffusion, or Dall-e, that can produce novel visual scenes from a text prompt (Islam 2022), there is now unprecedented capabilities for anyone to create 2D and 3D design content rapidly with ease. These advancements have led independent newspapers such as *The Guardian* to recently question whether the architectural profession will survive recent AI developments (Wainwright

2023). There is broad agreement that AI technologies are set to radically transform established models of life and work. In this precarious moment in time, although the possibilities that AI software might automate vast swathes of human labour gives cause for pause, there is an alternative to viewing the future as all doom and gloom.

Although popular media seems focused on predicting an "either-or" case for human operation versus level 10 autonomous systems, this book proffers an alternative model that strives for deeper partnership and engagement with semi-autonomous systems, suggesting a "both-and" model for design in the age of autonomy. The work presented embraces partnering with semi-autonomous systems. Specifically, the projects in this book demonstrate human critical and creative thinking operating in partnership with human-authored software and hardware. To date, this work has not engaged in large language model-based AIs. However, reflecting on recent developments, many of the projects presented could easily be extended by such AI technologies in several ways beginning with AI-assisted software development. Regardless of where in the spectrum of levels 1 to 10 of autonomy such developments could evolve towards, the author would be concerned with developing workflows that avert uncritical derivative work that comes easily from any technological advance. As will be evident in the book, the author engages in the development of technical methods within design to enable highly specific design outcomes that address human-led practical and aesthetic concerns.

Ethical Considerations:
In contrast to software and autonomous systems research taking place in other fields, the book's limited engagement with machine learning for task-specific training poses no risk of creating a rogue or super-intelligent AI or other dystopic scenario broadly speculated on at present. The work also did not involve the collection or use of people's data. As no determinations are made on people's behalf, the types of algorithmic bias exposed by computer scientists Joy Buolamwini and and Timnit Gebru are avoided (Buolamwini and Gebru 2018, 1). Nor is there any use of overly simplified heuristics for life-altering determinations that Virginia Eubanks and Cathy O'Neil suggest often fails to sufficiently represent or include the diversity of people within our communities, adversely impacting socio-economic equity (Eubanks 2019; O'Neil 2016).

Such unethical algorithms arise from what James Bridle describes as a *"universalist attitude"* where computation is considered the solution to all problems (Bridle 2019). In advocating for more degrees of autonomy in design, this book might be assumed to align with this attitude, but this is not the author's argument. In contrast to overly generalised algorithmic approaches, the book aims to engage very specifically and discretely with degrees of autonomy in partnership with people and other intuitive or deliberative activities on a case-by-case basis for creative and practical benefit. In this assisting or partnering role, event-based decision-making can support user-customisation, and design adaptation to site and environmental conditions, or other sensitive, non-presumptive, specific considerations. The book illustrates a variety of different approaches to design and aims to encourage designers to move towards more participatory frameworks; that engage with people/end-users, materials, manufacturing methods, software and AI.

Embracing degrees of autonomy in design is more of an opportunistic stance, that aims to leverage various technologies in combination with our own critical approaches without jettisoning our historical or cultural values. *In an industrialised world where artisanship has been marginalised by machinic processes and is no longer affordable or scalable to meet societal needs, this book seeks to bring creativity into the act of making.* This is not in opposition to our artisans or human labour, rather it supports an expansion in their capacity to engage in more variable, situated and scalable forms of artisanal impact, and as Chapter Two discusses, with greater ability to address socio-economic and environmental concerns. Although explorations in this book are often confined to a proof-of-concept, much of the work has the potential to be extended in partnership with craftspeople, and it is hoped that the work spurs new forms of human creative partnership.

Biomimicry
Although several biological references are made in the book, the author did not seek to mimic their operation or create aesthetic outcomes that appear natural. However, the systemic logic of computational research undertaken into natural systems has had some influence on the crafting of altogether different, bespoke design algorithms.

Process vs Artefact
Incorporating degrees of autonomy within design activities is not used to justify a design outcome simply due to the 'intelligence' of its process, or to remove the designer's engagement with aesthetics. On the contrary, design character is engaged proactively, and discussed frequently in the book. It is due to an interest in design outcomes that the book's design processes are not considered sacred but are continuously experimented on. The relationship established in this book between process and product is less related to architectural precedent than to molecular gastronomy.

Crafting Bespoke Designs through Autonomous Systems
While chefs such as Ferran Adrià (Jouary and Adrià 2014) or Heston Blumenthal (Blumenthal 2009) invent and iterate dishes through the employment of scientific methods and the development of custom techniques, their focus is on the multi-sensory experience of their dishes. Molecular gastronomy is extremely technical, but also utterly delightful! These chefs do not spend all of their creative hours in front of a sketchpad. Substantial time is spent in the kitchen, working with ingredients and tools, and interrogating preparation and cooking processes. Nothing is sacred. The fusion between technical and creative endeavours expands the degrees to which a dish's design can produce delight – taste might be accompanied by smell, sound, texture, or a surprising temporal event.

Architects and designers have similar opportunities when engaging with and orchestrating other agencies within project conception, development or realisation. This does not undermine their intuition or ability to intervene at any step within a design process. Intuition is extended into any activity that the designer has become accustomed to, and interventions are welcome wherever they are creatively or pragmatically productive. While the book's projects are provided as a limited set of examples, the opportunities for architects and designers are limitless. *Behavioural Production* explores a series of bespoke design approaches, crafted within diverse media. In expanding the media, methods and mode of thinking in which design takes place, the work suggests that the architect's creative role can partially shift to that of an initiator, mediator, event-organiser or co-pilot.

Multi-Agent Systems and Event-Based Design

Throughout this book, a particular approach to computation threads through the work. Custom-developed agent-based software programs are used to define spatiotemporal events where individual or collective agent behaviour produces unpredictable, emergent activity over time that is not explicitly described in encoded rules. Complex and coherent orders arise from processes of self-organisation that offer both problem-solving and aesthetic opportunities. The approach can be applied within 3D graphical space to design aspects of architecture in relation to several practical or creative considerations – essentially anything that might be computable. It can equally be deployed to create real-time adaptive behaviour in individual or collective robot systems. In this book, a particle, line, curve, surface, material, object, tool, industrial robot arm, mobile ground robot, or aerial robot may be programmed with design and/or manufacturing agency. Developed over more than two decades, this computational approach to behaviour-based design initially operated within 3D graphical space and was later expanded to include approaches to robotic fabrication and swarm-based, collective aerial additive manufacturing.

Synopsis: A Preview of Upcoming Chapters

Behavioural Production explores several approaches to event-based design, where various degrees of autonomous activity contribute to the resolution and expression of design outcomes. In each chapter's approaches to design, event-based rules are developed to co-opt diverse media's participation within the creative process, with the aim of expanding design agency and developing a bespoke design character. This approach is systemically open, adaptive to the specifics of each project and has the potential to shift the designers' perspective towards a more distanced, zoomed-out view of their role and the impact of their decisions. In this sense, the book offers a renewed interest in the aesthetic, objectified outcomes of design without negating the designer's responsibility to engage with other important considerations such as environmental or economic impacts.

Chapter Two explores the current state of the construction industry and its relationship to alarming societal and environmental concerns. The emerging Fourth Industrial Revolution's autonomous manufacturing and cyberphysical systems capabilities are also discussed, together with an overview of robotics in the construction industry and architectural robotics research being undertaken in academia. The work covered in this book is then differentiated from this large body of work by its emphasis on a behaviour-based computational design approach that engages with various media and scales of operation through degrees of autonomous programming, and in its production of behavioural character as an aesthetic condition. A scalable approach to design and production is also advocated for that is applicable to individual fabrication tasks through to the collective behaviour of robot swarm systems.

Chapter Three commences with anthropologist Lambros Malafouris' concept of *'material engagement'* and his example of a ceramicist working with a potter's wheel, where he argues creative outcomes are partly derived from agency in the clay and wheel. Material engagement can be strategised to extend design cognition to operate partly through other media and events. In this context, a series of exploratory projects are presented that leverage acts of manufacturing or assembly creatively. Designs are developed through the creative orchestration of manufacturing processes, where dynamic material processes of formation not only contribute to design outcomes but also strategically reduce material waste during production.

Chapter Four extends this generative design approach into computational space. An agent-based algorithmic design approach is advocated that embodies design intent within the semi-autonomous interactions of simulated point, curve and surface entities in 3D graphical space. A bespoke design character arises from the orchestration of behavioural events within these custom-developed, non-linear processes of self-organisation. The chapter explores a complex architecture where design intent is embedded within a negotiated, polyvalent architectural matter.

In Chapter Five, structural agency is incorporated into Chapter Four's algorithmic design approaches. Structural intent becomes embedded within design outcomes, as an inherent property of design character. In contrast to mathematically determined geometric approaches to structural design that use Gaussian curvature, graphic statics and other methods, a formative approach to structural design is presented more akin to the work of Antoni Gaudi, Heinz Isler or Frei Otto. While these pioneers produced structurally optimised forms, a non-optimal structural condition is explored that resolves structural considerations alongside other architectural criteria to produce negotiated design solutions. Unlike geometric design methods, the presented approach is readily extendable into non-geometric logistical and production concerns addressed in subsequent chapters.

Chapter Six introduces additive manufacturing (AM) technologies suited to building-scale work and some exciting possibilities afforded by an additively manufactured architecture. The chapter maps out a series of design considerations that seek to exploit AM to provide material and structural efficiencies to reduce material, cost, and environmental impact, while also expanding design expression through creative engagement with the manufacturing approach. In recognition of the locally differentiated and negotiated context in which most buildings arise, a multifarious, anisotropic, topological architecture is explored within a design approach that extends earlier chapters' computational methods to address AM architectural applications. A design methodology is developed for the topological formation of columns, walls, ceilings and structural bays, in addition to gravitational-material formation effects that can be controlled in material-extrusion (MEX) AM processes. A material-physics MEX simulation enables visual and structural analysis to be undertaken prior to manufacture, while some MEX-manufactured prototypical parts are also presented alongside computational design methods.

Chapter Seven explores several different approaches to incorporating autonomous perception in design. First in the analysis and generation of unique design character relative to other design and engineering criteria using computer vision and Deep Reinforcement Learning. Multiple approaches to generative design are also presented that leverage 3D site scanning and Deep Neural Networks (machine learning) within design workflows to explore a situated, bespoke site-tailored architecture. This specificity together with potentially rapid means of production also prompts speculation on an alternative form of architectural agency that operates under more spatiotemporal terms, and shares some relation to materialist, object-oriented and Posthuman discourses.

In Chapter Eight, a behaviour-based design approach to adaptive manufacturing is presented. Autonomous sensing and perception are incorporated into software that governs the behaviour of robot manufacturing tasks, recasting generative design as an embodied and situated process. In its assemblage of mass-produced parts, most present-day architecture arguably offers an over-generalised design solution. In contrast, this chapter's work theoretically supports the economical production of unique and considered design responses at industrial volumes of production. After demonstrating the approach at the scale of a material manufacturing process, the adaptive prefabrication of non-standard volumetric architectural modules is discussed, and the on-site activity of an autonomous mobile robot. The behaviour of mobile robots is then extended into adaptive approaches to autonomous aerial robot construction. Throughout these examples, degrees of spatiotemporal specificity and adaptation operate within design to produce a more distinctive design character that embodies vitality.

Chapter nine explores Collective Robotic Construction (CRC), expanding our capacity to design and build in unbounded and hard-to-access locations. Recently published as a cover article in Nature, the first-ever untethered additive manufacturing in-flight – Aerial AM is presented (Zhang et al. 2022), alongside architectural designs that emerge from the self-organising act of CRC. In contrast to an individual robot, the remote control of a robot population undertaking different concurrent activities is beyond the physical and cognitive capacity of an individual or team of people. Semi-autonomous programming is the sole means at our disposal to enable the robust coordination of aerial additive manufacturing in dynamic, outdoor environments. The chapter provides proof-of-concept in robot flight coordination, manufacturing and real-time collective design through physical robot flights, fabrication tasks, and simulation modelling. Some speculative architectural applications of the manufacturing method are also presented, designed during a simulated CRC process. CRC provides opportunities for site adaptive designs that might operate temporarily, or indefinitely as a process of continuous construction, producing a variable, sensitive, four-dimensional approach to architectural design. While still a nascent technology, CRC offers a scalable approach to design and construction that could potentially address projected levels of growth in the built environment that present-day technologies have no means to address.

The tenth, concluding chapter ties all of these endeavours together, advocating for the design of a highly specific architecture that goes beyond generalised design solutions that depend on Second and Third Industrial Revolution mass-production and mass-customisation methods. It suggests that Fourth Industrial Revolution autonomous manufacturing capabilities explored in this book can be more adaptive to the considerations of individuals, communities, and environments, enabling additional forms of design and production engagement with the world at large. The work presented in earlier chapters is shown to uniquely support a distributed and adaptive approach to manufacturing and construction, offering a scalable approach to bespoke design that arguably could be applied at industrial volumes of production.

Collectively, these chapters present a series of generative approaches to design that primarily operate through the orchestration of material, computational, or robotic events that involved extensive multidisciplinary collaborations. The book presents projects from the author's design practices – Kokkugia[1] and Robert Stuart-Smith Design[2]; the author's Autonomous Manufacturing Lab research laboratory in the University of Pennsylvania's Department of Architecture[3] and University College London's Department of Computer Science[4]; together with some research and teaching at the University of Pennsylvania[5] and Architectural Association School of Architecture's Design Research Laboratory[6]. Altogether, this experimental design research was developed within the context of a global shift in our means of thinking and operating that is taking place today. Two millennia after Heron of Alexandria's automata, the emerging Fourth Industrial Revolution (IR4) will have a significant societal and environmental impact. The next chapter discusses IR4 and its influence on architectural design and building construction.

Notes

1. *Kokkugia Ltd, 2004–2013. Co-Founding Directors Jonathan Podborsek, Roland Snooks, Robert Stuart-Smith.*

2. *Robert Stuart-Smith Design Ltd, 2013–Present. Co-Founding Directors Robert Stuart-Smith and Stella Dourtme.*

3. *Autonomous Manufacturing Lab research lab, University of Pennsylvania, Weitzman School of Design (AML-Penn). Director: Robert Stuart-Smith.*

4. *Autonomous Manufacturing Lab research lab, University College London, Department of Computer Science (AML-UCL). Directors: Robert Stuart-Smith, Vijay Pawar.*

5. *University of Pennsylvania design studios in MSD-RAS & M.Arch Program.*

6. *Studio Stuart-Smith, Architectural Association School of Architecture Design Research Laboratory.*

References

Alexandria, Hero, and Bennet Woodcroft. 2015. Pneumatica: The Pneumatics of Hero of Alexandria. CreateSpace Independent Publishing Platform.

"Autonomous." 2022. Merriam-Webster's Unabridged Dictionary. https://www.merriam-webster.com/dictionary/autonomous.

Blumenthal, Heston. 2009. The Fat Duck Cookbook. Bloomsbury USA.

Bonabeau, Eric, Sylvain Guérin, Dominique Snyers, Pascale Kuntz, and Guy Theraulaz. 2000. "Three-Dimensional Architectures Grown by Simple 'stigmergic' Agents." BioSystems 56 (1): 13–32. doi:10.1016/S0303-2647(00)00067-8.

Braitenberg, Valentino. 1986. Vehicles: Experiments in Synthetic Psychology. Bradford Books. MIT Press.

Bridle, James. 2019. New Dark Age: Technology and the End of the Future. Verso Books.

Buolamwini, Joy, and Timnit Gebru. 2018. "Gender Shades: Intersectional Accuracy Disparities in Commercial Gender Classification." In Proceedings of the 1st Conference on Fairness, Accountability and Transparency, edited by Sorelle A Friedler and Christo Wilson, 81:77–91. Proceedings of Machine Learning Research. PMLR. https://proceedings.mlr.press/v81/buolamwini18a.html.

Butler, Samuel. 1970. The Iliad Of Homer: Samuel Butler's All Time Bestseller Classic. Kindle iOS version.

Cartwright, Mark. 2019. "Hephaistos." World History Encyclopedia. September 3. https://www.worldhistory.org/Hephaistos/.

Dorigo, Marco, Eric Bonabeau, and Guy Theraulaz. 2000. "Ant Algorithms and Stigmergy." Future Generation Computer Systems 16 (8). Elsevier: 851–71.

Eubanks, Virginia. 2019. Automating Inequality: How High-Tech Tools Profile, Police, and Punish the Poor. Picador.

Grassé, Pierre-Paul. 1959. "La Reconstruction Du Nid et Les Coordinations Interindividuelles Chez Bellicositermes Natalensis et Cubitermes Sp. La Théorie de La Stigmergie: Essai d'interprétation Du Comportement Des Termites Constructeurs." Insectes Sociaux 6 (1). doi:10.1007/BF02223791.

Grillo, Francesco. 2019. "Hero of Alexandria's Automata: A Critical Edition and Translation, Including a Commentary on Book One." Glasgow.

Henderson, Georgina J. 2019. "Gods and Robots: Myths, Machines, and Ancient Dreams of Technology by Adrienne Mayor." Technology and Culture 60 (4). doi:10.1353/tech.2019.0117.

alJazari, Isma'il ibn al-Razzaz, and Donald Routledge Hill. 1975. The Book of Knowledge of Ingenious Mechanical Devices: Kitab Fi Ma Rifat Al-Hiyal Al-Handasiyya. Springer Netherlands.

Islam, Arham. 2022. "How Do DALL·E 2, Stable Diffusion, and Midjourney Work?" MarkTechPost. November 14. https://www.marktechpost.com/2022/11/14/how-do-dall%C2%B7e-2-stable-diffusion-and-midjourney-work/.

Jordan, John M. 2016. Robots. The MIT Press Essential Knowledge Series. MIT Press.

Jouary, Jean-Paul, and Ferran Adrià. 2014. Ferran Adria and ElBulli: The Art, The Philosophy, The Gastronomy. Harry N. Abrams.

Kotsanas, Kostas. 2018. Ancient Greek Technology: The Inventions of the Ancient Greeks. Athens: Kotsanas Museum of Ancient Greek Technology.

Krishna, Kumar, and David A Grimaldi. 2003. "The First Cretaceous Rhinotermitidae (Isoptera): A New Species, Genus, and Subfamily in Burmese Amber." American Museum Novitates 2003 (3390). BioOne: 1–10.

Matarić, Maja J. 2007. The Robotics Primer. Intelligent Robotics and Autonomous Agents. Cambridge University Press.

Merriam-Webster. 2023. "Behavior Definition & Meaning – Merriam-Webster." Merriam-Webster Dictionary. June 1. https://www.merriam-webster.com/dictionary/behavior.

O'Neil, Cathy. 2016. Weapons of Math Destruction: How Big Data Increases Inequality and Threatens Democracy. New York: Crown.

OpenAI. 2023. "GPT-4." March 14. https://openai.com/research/gpt-4.

Pask, Gordon, and Susan Curran. 1982. Micro Man: Computers and the Evolution of Consciousness. Macmillan.

Reynolds, Craig W. 1987. "Flocks, Herds, and Schools: A Distributed Behavioral Model." In Proceedings of the 14th Annual Conference on Computer Graphics and Interactive Techniques, SIGGRAPH 1987, 25–34. Association for Computing Machinery, Inc. doi:10.1145/37401.37406.

Schwartz, Jacob T, John von Neumann, and Arthur W Burks. 1967. "Theory of Self-Reproducing Automata." Mathematics of Computation 21 (100). doi:10.2307/2005041.

Sharkey, Noel. 2007. "I Ropebot." New Scientist 195 (2611).

Sheridan, Thomas B, and William L. Verplank. 1978. "Human and Computer Control of Undersea Teleoperators." ManMachine Systems Lab Department of Mechanical Engineering MIT Grant N0001477C0256. doi:10.1080/02724634.1993.10011505.

Wainwright, Oliver. 2023. "'It's Already Way beyond What Humans Can Do': Will AI Wipe out Architects?" The Guardian. August 7. https://www.theguardian.com/artanddesign/2023/aug/07/ai-architects-revolutionising-corbusier-architecture?CMP=Share_iOSApp_Other.

Walter, William Grey. 1953. The Living Brain. Norton.

Wolfram, Stephen. 2002. A New Kind of Science. General Science. Wolfram Media.

Wong, Jenny, and Lisa Hendry. 2023. "The Origin of Our Species." The Natural History Museum. Accessed May 31. https://www.nhm.ac.uk/discover/the-origin-of-our-species.html.

Zhang, Ketao, Pisak Chermprayong, Feng Xiao, Dimos Tzoumanikas, Barrie Dams, Sebastian Kay, Basaran Bahadir Kocer, Alec Burns, Lachlan Orr, Talib Alhinai, Christopher Choi, Durgesh Dattatray Darekar, Wenbin Li, Steven Hirschmann, Valentina Soana, Shamsiah Awang Ngah, Clément Grillot, Sina Sareh, Ashutosh Choubey, Laura Margheri, Vijay M Pawar, Richard J Ball, Chris Williams, Paul Shepherd, Stefan Leutenegger, Robert Stuart-Smith & Mirko Kovac. 2022. "Aerial Additive Manufacturing with Multiple Autonomous Robots." Nature 609 (7928): 709–17. doi:10.1038/s41586-022-04988-4.

This chapter provides a context for the generative design research outlined in this book, demonstrating a socio-economic, environmental and industry need to leverage Fourth Industrial Revolution (IR4) technologies to enable more rapid, materially efficient, site- and user-specific, and creatively unique architectural designs. An opportunity to expand creative design expression through IR4 autonomous technologies is also introduced that connects to algorithmic and material research described in Chapters Three to Five and continues into robotics-based work covered in Chapters Six to Nine.

Our Constructed Present is Inadequate. A Band-Aid Won't Do
Buildings play a vital humanitarian and societal role, providing shelter and security, and catering for specific needs across diverse building types from housing to airports, office buildings, hospitals, schools and many others. However, the economic and environmental cost of this essential aspect of society is staggering.

An Unsustainable Built Environment
In early 2020, the *New York Times* ran a story about a subsidised 'Affordable Housing' development in San Francisco which had a build (not sale) cost of $750,000 per apartment – more than three times the average sale price of a family home in the US despite receiving several subsidies (Fuller 2020). With costs as ludicrous as this, current building practices are not only inadequate to support equitable home ownership possibilities, but are also incapable of providing sufficient housing for those in need. The *New York Times* estimated that for California to build 150,000 homes to house all of California's homeless population, it would cost $70 billion, a number so astronomically high it is unlikely to ever happen (Fuller 2020). Across the Atlantic Ocean, the UK, predicted to be the sixth largest construction market in the world by 2030 (Betts et al. 2015, 15), is in need of an additional 4.75 million homes. Current government initiatives are projected to only reduce this by one-quarter over the next ten years while demand will continue to increase (Wilson and Barton 2023, 15). In 2015 it was estimated that India, the world's largest housing market, would have to build 170 million homes within 15 years to house its anticipated urban population in 2030 (Betts et al. 2015, 15). This present condition foreshadows an even more challenging future; the world's population is expected to exceed 9.7 billion by 2050 (UN DESA 2014). With growing rural to urban migration, there

a

b

c

d

e

f

is a critical need to provide buildings to support an estimated 2.5 billion additional people living and working in urban populations by 2050 (UN DESA, 2014). Present practices will not be able to meet the levels of productivity and affordability required to overcome this challenge.

The environmental cost of building is also extremely high, with buildings accounting for 39% of energy, 40% of CO_2 emissions, and 40% of raw material use each year (Bergman 2013, 15,24–25,62). Part of this material consumption is simply due to wasteful practices. In the UK, 10% of new materials delivered to a construction site go directly to a landfill, while 30% of total building material weight is wasted in the US. In the EU, 75% of construction waste isn't even recycled (Osmani 2011, 209). Given that building supply will need to increase significantly to support projected urban population growth, current construction methods will only further hasten climate change and environmental damage. Reducing material use, waste, and energy consumption in both buildings and construction is not only beneficial to the environment, these savings would also reduce the cost of buildings, providing a win-win scenario for the environment, inhabitants, governments and developers. To satisfy increasing demand, a faster and more flexible approach to construction is required, that can be adapted to the diverse scenarios in which buildings are needed. Unfortunately, the global construction sector is endemically unproductive. Although building supply needs are not equally distributed across all countries, small emerging economies through to large established ones share the common need to increase supply and productivity. McKinsey's research suggests that countries with high building demand and high labour costs[1] also stand to benefit greatly by incorporating robotics and automation (McKinsey Global Institute 2017, 11–12). In the UK, the US and several other economies, not only are these conditions met, but the construction sector is also experiencing a decades-long trend in workforce decline.

Construction in Crisis

2.1 Built Environment
a) Construction productivity is stagnant or in decline while other manufacturing sectors have increased productivity over decades, b) high number of fatalities and injuries, c) skilled labour decline, d) high material consumption and waste, e) costly, not affordable, f) large environmental impact.

In the UK, anticipated growth in building activity is contrasted by an anticipated 20–25% decline in the construction labour force over this decade due to an ageing workforce (Farmer 2016, 8), a problem also shared with the US (Janicki and McEntarfer 2015). This is perhaps partly attributed to the unfavourable conditions in which construction workers operate. In the UK, it is a dangerous industry to work in, with double the fatalities of all other manufacturing activities combined (HSE 2021, 6). In the United States, statistics are equally grim, with construction worker fatalities almost three times greater than in other manufacturing sectors and trending upwards (Bureau of Labor Statistics 2021). Fatalities are estimated to be 3–6 times worse in developing economies (International Labour Organization 2015). Beyond lethal, in many leading economies, construction is far from productive.

The global construction sector's productivity growth is abismal relative to economic and manufacturing growth over the last two decades. In the US, where agriculture and manufacturing sectors have expanded productivity by 10–15 times since the 1950s, the US construction sector's productivity has barely changed (McKinsey Global Institute 2017, 1). Representing 13% of the world's GDP, this lack of productivity embodies a significant stagnation and economic loss (McKinsey Global Institute 2017, iii). Although it is projected to become the world's sixth-largest construction market by 2030 (Betts et al. 2015, 8), the UK construction sector has been the least productive sector of the UK economy for several decades. A series of UK government reports have identified problems including low productivity and predictability along with low margins causing financial fragility. Low levels of vertical integration in the construction supply chain and a high reliance on sub-contracting were also believed to fracture design, construction management, and building execution, leading to high construction costs and low levels of innovation or investment in research and development (Farmer 2016; HM Government 2013). Delays to a project's completion also create astronomical financial losses for building owners and construction companies. While UK low-rise buildings are mostly completed on schedule, only a third of high-rise buildings are. With at least 18% completed more than six months late, the industry is struggling to effectively scale up operations to suit larger-sized projects (Farmer 2016, 16). Established building practices are failing to produce enough buildings, or as rapidly, predictably, and as cost-effectively as society needs. Unfortunately, these issues cannot be resolved by making minor adjustments to the way buildings are designed or constructed. Radical and systemic change is required to overcome endemic low productivity and the significant environmental and economic costs of building.

2.2 Urban Growth
People are increasingly migrating to urban environments. With projected population growth, there is an astronomical amount of urban building that must be undertaken by 2050 to meet demand.

Towards Autonomous Construction

Taking a more proactive outlook, the industry's lack of productivity presents a great opportunity. Another UK government report concluded that if construction leveraged digital and robotic technologies it could drive an additional £88.9bn in economic growth over ten years, also generating £3bn in savings passed on to consumers and 365,000 tCO$_2$e reductions, while saving 105 lives through improved worker health & safety. Increases in living standards, quality of construction, and consumer and worker satisfaction were also identified (Maier, 2017). Although this exciting forecast is yet to be realised, in 2018 the UK government initiated the *'Transforming Construction Challenge' (TCC),* providing a combined £420 million from government and industry to support projects that *"accelerate a shift in construction...towards manufacturing and digital processes"* (some TCC-funded research is presented in Chapter Eight) (UKRI, 2021). The benefits of automating construction are also not exclusive to G8 economies operating with high labour costs. Research has indicated that introducing robots into India's construction industry for specific tasks would result in a 60% reduction in time, cost and rework compared to current manual labour practices (Kumar, Balasubramanian, and Raj 2016). Although this research suggests potential, leveraging digital and robotic technologies effectively in building construction is challenging. A more integrated approach to the design and construction of buildings is required. Fortunately, a major upgrade to the manufacturing industries has commenced that will forever impact the built environment. It is transformative enough to have been labelled as the Fourth Industrial Revolution.

Industry 4.0 (IR4)

Industrial revolutions are named such due to the significant boost they have provided in technological advancement, together with various undesirable side-effects that have irreversibly impacted social, economic, geopolitical and environmental landscapes. The first industrial revolution (IR1) was based on advances in mechanical production and steam power during the eighteenth century and was followed by a second industrial revolution (IR2) in the nineteenth century that ushered in mass production and electrical energy. The third (IR3) leveraged information technology to bring advances in electronics and the emergence of the Internet. So, what is the Fourth Industrial Revolution (IR4)? Sometimes referred to as Industry 4.0, it describes a shift to a *reliance on cyber-physical systems and autonomous manufacturing.* Although this may seem like science fiction, it isn't. Several of the world's governments are dedicating significant research and development funding to spur economic growth from IR4 technologies to ensure they can compete in this fast-paced and fundamental shift in the global market. According to Tobias Schwab, economist and founder of the World Economic Forum, IR4 has four main physical manifestations in the short term; autonomous vehicles, 3D printing, advanced robotics, and new materials (Schwab 2017). All of these are already having a direct impact on the building industry, with all four discussed to some degree in this book.

Advancing Manufacturing and Construction

While autonomous vehicles and 3D printing are clearly emerging technologies, robots have been in factories for decades performing automated tasks. In IR4 though, a new generation of robots will operate more autonomously. By operating as cyber-physical systems – that integrate activities such as computation, communication, sensing and networking within physical objects or infrastructure (NSF 2023) – IR4 robots will share learning experiences, adapting quickly while supporting greater collaborations between people and robots. Leveraging these capabilities within the construction industry, however, requires an upgrade to how buildings are digitally modelled.

IR3 technologies enabled the building industry to digitally coordinate construction documents and activities around a Building Information Model (BIM) that combined architectural, structural, mechanical and electrical information, and supported file-to-factory workflows. IR4 goes one step further, using a 'digital twin'. Initially developed in NASA's 1960s space program, a digital twin operates as a multi-physics simulation that mirrors the life of a physical vehicle, system (or building) throughout its use, but it can also be used in concept design to debug a design's operation prior to fabrication (Shafto et al. 2010, TA11-7; Grieves and Vickers 2016). An IR4 building digital twin could be used to track, calibrate, simulate, or communicate with real-world fixed assets such as buildings or building parts, together with mobile assets such as robots, vehicles, clothing, tools, and electrical devices, in addition to people and goods. Leveraging communication and location-tracking technologies together with computer vision and sensor systems, it is possible to analyse, predict, improve, or adapt building or manufacturing tasks by feeding back real-world data. While this raises ethical concerns around privacy that must be addressed, there is immense potential to transform the supply, demand and distribution of building materials, services, and manufacturing, and to automate more construction activities. A digital twin could also model time estimates and logistics to provide greater certainty, reducing risks and time delays. Coupling a digital twin with increasingly sophisticated means of robotic manufacturing will also enable higher-quality buildings to be more precisely built with less material, time and cost while supporting diverse design possibilities.

How might architectural design leverage IR4 technologies?
IR4's technological impact on design is best understood relative to IR2 and IR3 possibilities. IR2 enabled the economical mass-production of modular prefabricated building parts which led to tectonic and aesthetic possibilities, particularly in steel, concrete and glass material systems. While works that express prefabricated modular elements such as structural elements in Paxton's Crystal Palace or late twentieth-century CE British High-Tech architecture fall into this category, Europe's post-war high-density housing developments offer perhaps the best example of IR2 architecture's merits and limitations. Modernist designs comprised of modular parts offered an economical approach to higher-quality housing for the masses and could more readily support rapid construction efforts in comparison to traditional building techniques. However, the economics and logistical constraints of IR2 production also encouraged geometrically repetitive design solutions that privileged a systemic approach to parts, arguably at the expense of some flexibility in a building's overall design. Some projects were even repeated on several sites – effectively representing an ideology of mass production for both building parts and buildings. Le Corbusier's Unité d'Habitation was constructed in different locations, while Mies van der Rohe's several iterations of a tower design typology that includes his New York Seagram Building and One IBM Plaza in Chicago amongst several others offer clear examples of this. Such approaches arguably generalise design considerations to suit the manufacturing paradigm's systemic limitations.

IR3's development of information technologies gave rise to the use of computer algorithms, 3D parametric models and BIM within architectural design and documentation that enabled architects to easily develop projects that embodied geometric variation within a building's constituent parts. The digital media in which these professional services took place also supported file-to-factory pipelines, aligning with IR3 mass-customisation production methods. The benefits of IR3 mass-customisation is most evident in building designs that exhibit degrees of visual difference in the tiling or organisation of parts across a façade or structure, or in their volumetric/formal variation. While mass-customisation enables a departure from the limitations of IR2 mass-production methods, it typically costs more, resulting in IR2 methods remaining the dominant means of building production today. This cost difference decreases with economies of scale. For this reason, designs leveraging IR3 technologies are primarily achieved in larger-scale, high-budget projects such as Frank Gehry's Guggenheim Museum in Bilbao (Brayer 2015), or Herzog DeMeuron's 2008 Olympic Stadium in Beijing (Webb 2008). Economical differences can be partially offset by limiting the variation of parts to a lesser identifiable quantity as is evident in the same olympic's National Acquatic Centre (also known as the "Water Cube"), where a Voronoi façade was designed by PTW and Arup to appear differentiated but uses a small set of repeated geometrical types to limit the number of different elements and nodal connection types (Etherington 2008). Similarly, a Penrose fractal tiling was utilised on façades in Lab Architecture Studio's Federation Square project to provide the appearance of more variation than its limited number of material and panel types supported (Day et al. 2005).

While IR3 architectural outcomes rely on narrow margins of variation within established products and systems, IR4 autonomous manufacturing can theoretically support bespoke, one-off designs at industrial levels of production. This is radical. Although historically, bespoke designs were exclusively available to elite tiers of society, IR4 manufacturing methods have the potential to enable all buildings to be realised as one-off designs without necessarily incurring time or cost penalties (additive manufacturing provides a clear example of this, discussed in Chapter Six). Where architects operating in IR2 and IR3 paradigms had to partially generalise a design problem to suit industrial modes of production, IR4 enables an increased degree of variability in some autonomous manufacturing processes that can support more distinctive and considered design responses that can challenge spatial, formal, tectonic and material orders, and more, at multiple scales. This can lead to greater specificity in designs, that provide more effective means to address local or regional differences in climate, topography, urbanism, social, cultural, and economic conditions in addition to individual client or end-user's needs. IR4 will also enable new opportunities for near-instant personalised services and experiences, enabling architecture-in-use to be more sensitive to a diversity of people's needs and aspirations. While most of today's architectural designs are arguably generalised solutions, constrained by their assemblage of IR2 mass-produced and IR3 mass-customised parts, IR4 enables each building design to embody a uniquely considered response (this doesn't preclude IR4 methods being used in conjunction with IR2 and IR3 approaches). But how far away is the building industry from implementing IR4 methods? While it has not kept pace with many other industries and requires much more development, the construction industry has made strides in this direction.

Towards Automating Construction

Task-Specific Building Robots
Since the early 1970s, several Japanese construction companies[2] have utilised a vast range of task-specific robots for on-site construction activities including: steel reinforcement bending and installation, brick laying, concrete pouring using fixed and mobile robots, mobile concrete compaction and finishing, welding, heavy lifting, and robots dedicated to the assembly of façade glazing panels, tiling, plasterboard sheets, along with painting, and more (Bock and Linner 2016a, 45, 75, 81–83, 105, 148, 44, 164, 177, 216, 185). Some of these robots provide massive increases in productivity. In the US, a series of mobile robots now offer promising construction capabilities including Boston Dynamics quadruped *Spot™* robot's deployment for on-site data collection (Trimble 2022) and Dusty Robotics *FieldPrinter™* wheeled robot that prints out site-construction layout information on floor slabs (Dusty Robotics 2022). Across the other side of the Pacific Ocean, Australia's Fastbrick Robotics™ brick-laying robot *Hadrian X™* is able to build the walls of a house in one day at a rate of one thousand bricks per hour, with almost zero waste, more rapidly, accurately, and cost-effectively than traditional methods (FBR Ltd 2018). In China, Guangdong Bozhilin Robot Co Ltd™ now rents, sells and offers services in robotic construction for building and finishing operations. In their first three years of operations, the company helped build 120 projects using eighteen different self-developed robot platforms (iNEWS 2022). In earthworks and road construction, Xuzhou Construction Machinery Group™ recently used several robot construction rollers and paving machines to resurface a stretch of Nanjing–Shanghai Expressway (Rogers 2021), while Baidu Research Robotics and Auto Driving Lab (RAL) at the University of Maryland has demonstrated an autonomous excavator robot continuously excavating soil for 24hrs at a comparable rate to a human-operated vehicle (GCR Staff 2021). US Company Autonomous Solutions Inc (ASI™) has also developed robot hardware that converts human-operated plant equipment such as dump trucks and excavators into autonomous mobile systems (Bock and Linner 2016a, 36). Collectively these examples demonstrate the existence of a broad range of robotic platforms and capabilities for different construction activities.

Off-Site Manufacturing
While task-specific robots can increase the production efficiency of on-site operations, construction productivity could improve by 60% simply by undertaking at least 70% of activities off-site, where factory-like conditions provide greater degrees of safety and control and are not disrupted by weather or traffic (Science and Technology Select Committee 2018). Laing O'Rourke successfully built 80% of London's 47-storey Leadenhall Building off-site. Ship- and bridge-building techniques inspired the prefabrication of the primary steel structure into sections that each incorporated seven floors. The ambition of the project required two custom cranes to be made that could lift forty-five tonne payloads. The approach demonstrated improvements in build quality, safety and speed, and proved to be less disruptive and more streamlined in assembly operations – invaluable on such a busy and constrained site in central London (Leadenhall Building 2023). Developers Forest City Ratner together with construction company Skanska and Shop Architects were similarly ambitious in building the world's first high-rise modular tower, 461 Dean St in Brooklyn, New York in 2016. At 32 stories, the residential tower stacked 930 self-structured

modular prefabricated steel spatial units on top of one another like Lego, reducing on-site work to primarily excavation and foundation construction activities (Touhey 2016). Although these projects were not primarily manufactured by robots, US-based companies Blueprint Robotics™ and Autovol™ both provide commercial off-site manufacturing services that are heavily invested in robotic prefabrication of building parts for multi-storey buildings with hundreds of apartments (Autovol 2022; Blueprint Robotics 2023). It doesn't take much of a leap to connect their workflows to projects of similar scale as Leadenhall to bring further time and productivity benefits. Canadian Firm Intelligent City already has such capabilities for mass-timber prefabrication (Intelligent City 2023).

On-Site Automation

As an alternative to off-site building manufacturing, several Japanese construction companies including Obayashi™ and Shimizu™ have turned some of their construction sites into giant *'Sky Factories'* that move upwards in step with building progress. Far from a one-off experiment, they have developed and deployed over thirty different site factory platforms on more than sixty high-rise projects (Bock and Linner 2016b, 1). A gantry platform large enough to enclose a whole floor is utilised throughout construction. Similar to jump-form concrete construction, after building work is completed on one floor, the platform is shifted upwards to rest on its previous work while building the next floor. Compared with these companies' other project sites, Sky Factory sites delivered substantial increases in construction speed, doubling productivity despite using 20–70% less human labour, reducing waste by more than 70%, and work-related injuries (Bock and Linner 2016b, 303). Interest in the method is not limited to Japan. UK construction company Mace has developed a similar system called *'Jump Factory'*, which it claims will make construction six times more productive than current UK methods (Gerrard 2019). While current Sky Factories are more mechanistic than robotic, they could easily be further roboticised in their operations and incorporate just-in-time logistics similar to those utilised by Japanese car manufacturers. A "Lean Manufacturing" approach could strategise material and product supply chains, labour, transport and other logistics, all of which are easily quantifiable and interconnected using IR4 technologies (Weiss 2013).

Constraints to Design Creativity

Sky Factories and task-specific robot systems primarily address the automation of existing IR2 and IR3 building methods. Compared to an IR4 approach to autonomous manufacturing, they are greatly limited in their ability to manufacture variable design outcomes. While a task-specific robot plasterer could vary the dimensions of its work, a robotic window installer can only install prefabricated window units (most likely built by IR2/3 methods). Sky Factories use of previously built floors as a support structure throughout the build process likely constrains building designs to simple vertically extruded geometries that share a common plan outline shape. Although these industry developments point towards a more automated construction sector capable of delivering projects at greater speeds, it is not clear how design might be able to leverage these technologies. Notably, their easily repeatable workflows potentially encourage the continued use of IR2 mass-produced prefabricated parts. An alternative robotic manufacturing and construction method – Additive Manufacturing – is perhaps more suggestive of how design could leverage IR4 logistics to manufacture one-off bespoke designs, due to its agnosticism to geometry and ability to be used at the scale of a part or whole building (see Chapter Six). Additionally, academic research institutes have been exploring the design possibilities of robotic fabrication and construction for quite some time, demonstrating exciting possibilities for using robot systems to positively impact architectural design alongside manufacturing and construction activities.

Design Research Leveraging Robotic Construction

As early as 1967, Marvin Minsky developed a robot arm that could build with toy blocks using computer vision (Minsky 2011). Three years later, in the exhibition '*Software*' in New York, MIT's Architecture Machine Group led by Nicholas Negroponte, showcased an art installation called '*SEEK*' that incorporated a robot gantry within a gerbil habitat, demonstrating a continuously adaptive form of built environment (Architecture Machine Group MIT 1970). More recently, starting with Gramazio Kohler's research at ETHZ Switzerland in 2005 (Gramazio, Kohler, and Willmann 2014), and Achim Menges' pavilion designs soon after at ICD University of Stuttgart (Menges and Knippers 2011), six-axis industrial (articulated) robot arms[3] have been utilised in an increasing amount of architectural design and fabrication research. Primarily used for automated manufacturing tasks, in 2020, three million industrial robots were estimated to be in operation globally, with a yearly growth trend of 14% (Spong, Hutchinson, and Vidyasagar 2020, 3).

There is now substantial research and teaching activity in robotic fabrication being undertaken across a vast number of architectural institutions internationally, exploring a wide range of building materials and fabrication methods. Examples include the variable placement and orientation of bricks or timber studs into geometrically irregular assemblages (Seo, Yim, and Kumar 2013; Doerfler et al. 2016; Wu et al. 2022), timber shell structures (Menges and Knippers 2011), woven fibre-composite pavilions (Doerstelmann et al. 2015), incrementally formed sheet metal structures (Nicholas et al. 2020) water-jet cut façade screens (King and Grinham 2013), self-structuring pavilions (Barandy 2018), bespoke cut stone (Clifford, McGee, and Muhonen 2018) and timber walls (Jahn, Wit, and Pazzi 2019) or metal rod bending installations (Pigram and McGee 2011), and a locally sourced tree-trunk trussed roof (Mairs 2016). Precast and in-situ concrete part and shell designs have been realised using robot-woven fabric formwork (van Mele et al. 2019), robot hot-wire cut EPS foam (Feringa, McGee, and Søndergaard 2013; Rust et al. 2016) or by large-format additive manufacturing (Jipa et al. 2016; Meibodi et al. 2018). Industrial robot additive manufacturing has also been explored in concrete (Anton et al. 2020; Bhooshan, van Mele, and Block 2020; Buswell et al. 2007), metal (Buchanan and Gardner 2019; MX3D 2020), ceramics (Rael and Fratello 2017; Bechthold, Kane, and King 2015; Sabin 2010), plastics (Soler, Retsin, and Garcia 2017), composites (Mohamed, Bao, and Snooks 2021), and expanding foam (Keating et al. 2017), alongside more novel construction methods such as string-rock jammed structures (Aejmelaeus-Lindström et al. 2016).

Additionally, aerial robot construction of a tensile bridge (Augugliaro et al. 2013), tensile installations (Braithwaite et al. 2018; Stuart-Smith 2016), assembly of a foam block scale-model building (Augugliaro et al. 2014) and aerial additive manufacturing in-flight (Zhang et al. 2022) have been demonstrated. Custom building robots have also been developed that walk (Jones 2015), climb walls (Yablonina et al. 2017) or build on top of previously built material (Jokic et al. 2022; Kayser et al. 2019; Petersen, Nagpal, and Werfel 2011) or modular robots that form adaptive building structures (Galloway, Jois, and Yim 2010). The scientific community has also developed a substantial body of research in custom robot platforms for collective robotic construction (Petersen et al. 2019), and there is an expanding research community exploring mixed-reality interfaces for human-robot interactive systems (Johns and Anderson 2018; Hahm 2019; Jahn, Wit, and Pazzi 2019) and forms of real-time robot programming (Castillo Y López 2019).

Although most architectural robotics research has not gained traction in the building industry to date, collectively, the work demonstrates a significant technical and design capability in a diverse range of automated manufacturing methods that have been extensively published in journals and conferences. Countless books also document this research in great detail, including *The Robot Touch: How Robots Change Architecture* (Gramazio, Kohler, and Willmann 2014), *Digital Fabrication* (Yuan, Leach, and Menges 2018), *Towards a Robotic Architecture* (Daas and Wit 2018), and numerous *AD: Architectural Design* editions (Menges 2015a; Tibbits 2017) amongst others. While many of the exceptional research examples mentioned highlight design opportunities in automating individual manufacturing or construction applications using robotics, they represent a multitude of approaches to computation, design, materials, and manufacturing methods. It is difficult to parse an approach to design that might accommodate such diverse activities within a sufficiently inclusive conceptual framework or envisage a trajectory for this work that fully leverages emerging IR4 technology developments.

An Explorative Framework for Creative Autonomous Manufacturing
In the building industry, robots are generally used as automated manufacturing tools, programmed to perform a predefined task in a similar fashion to a CNC machine. Notwithstanding a small number of exceptions, most previously mentioned research examples programmed automated robot routines to perform narrowly defined, predetermined tasks. Conversely, IR4 is enabling a shift towards highly capable autonomous systems. Supported by an emerging cyber-physical infrastructure, even current robot systems could be programmed to perform various tasks autonomously. There are opportunities, therefore, for degrees of autonomous activity to be incorporated within architectural design intent. In some cases, robots can contribute to design through the act of making. However, to instruct robots to behave more creatively, a creative approach to their programming is required. In the cyber-physical environment of IR4, robot programming can be extended to encompass degrees of autonomous robot control, that might govern a robot's interactions with dynamic material or environmental conditions.

In *Authoring Robotic Processes*, Gramazio, Kohler and Willman suggested architecture was at the *"dawn of a second digital age"* where through robot systems, computational methods could inform material processes, arguing that a *"digital materiality"* offered new aesthetic and functional possibilities (Fabio Gramazio, Kohler, and Willmann 2014, 14). Achim Menges has also argued that emerging cyber-physical production capabilities challenge architecture's established approaches to explicit geometric design and production, and enable generatively sensed, processed and adaptive forms of *"convergence between design and materialisation"* (Menges 2015b). The work explored in this book is significantly aligned to Menges' notion of adaptive design that is primarily operating outside of a priori, geometric or automated approaches, but it is also deeply engaged in the inherent aesthetic implications of such processes as Gramazio, Kohler and Willmann highlighted. More specifically, however, this book documents aesthetic outcomes derived from custom-developed behaviour-based systems. Commenced more than two decades ago, this work explores the production of behavioural character within algorithmic, material, robotic and collective robotic systems.

While Gramazio and Kohler's or Menges' work operates within robotic activities that utilise a range of parametric, algorithmic, and behavioural approaches, the work covered in this book is primarily situated in a behaviour-based computational design approach that engages with various media and scales of operation through degrees of autonomous programming. Across these works, the production of behavioural character as an aesthetic condition is explored. This focused computational approach towards autonomous behaviour would theoretically support the entire set of design approaches documented in this book to be connected into a single mega-methodology if one so wished. As a primarily bottom-up approach to design that engages in semi-autonomous processes, the work also offers a scalable approach to design and production that operates from individual fabrication tasks through to the collective behaviour of robot swarm systems. Aside from previously mentioned exceptions, most architectural robotics research to date has involved the programming of individual robots (or single multi-robot work cells) to execute automated routines. In contrast, the semi-autonomous approach advocated for in this book is, in principle, more readily able to engage variable site, scale, design, fabrication or construction scenarios with a flexible number of robots. Arguably, therefore, this book at least conceptually, suggests a means to address the vast socio-economic and environmental challenges mentioned at the outset of this chapter, that are impractical to resolve through current automated practices that can't scale to higher volumes or locations of production.

Compared to the diverse material and manufacturing methods explored in the broader academic community's architectural robotics research, the work covered in this book operates mostly within computational methods and their relation to individual robot and swarm-based approaches to additive manufacturing. The book explores a behavioural approach to design that commenced around 2002 with the development of generative algorithmic design approaches and subsequently evolved in more recent years to incorporate a similar attitude towards materials, robot systems, and their sensing and tooling capabilities. Inspired by computer scientist Craig Reynolds' behavioural approach to software programming (Reynolds 1999), and robotics pioneer Rodney Brooks' 'Subsumption Architecture' (Brooks 1985), a behaviour-based approach to algorithmic design is presented that operates in both computational and physical environments to engage in the materialisation of emergent design outcomes.

Although this work is directed towards both practical and creative goals, it also remains exploratory. The specific orchestration of autonomous events within a project might, for instance, provide space to challenge established design or production assumptions. Leveraging semi-autonomous means of perception, sensation, and motion – design and production possibilities can extend beyond human physical or cognitive capabilities. This does not suggest designers should leave their brains or bodies behind, but rather that through creative approaches to robot programming, designers can discover additional forms of agency in robot bodies, their sensing and interaction with materials, forces, and effects, or through their collective self-organising action. Bespoke design character is thus also explored through the specific design of a tool, the particular physical capabilities of a robot platform or through several computer agents' or robots' collective behaviour. In this sense, the work is at times opportunistic or speculative. While this expands design activities into diverse media and circumstances, it requires a designer to realise their design intent while relinquishing some direct control to various degrees of autonomous behaviour.

Behavioural Production: Semi-Autonomous Approaches to Architectural Design, Robotic Fabrication and Collective Robotic Construction explores an engagement with materials, fabrication, construction, computation, robotics and collective robotic systems as valued participants in design conception and realisation. Throughout most of the work presented in this book, design operates primarily within generative algorithmic approaches to behaviour. Many such approaches also mediate and inform material and manufacturing activities. Design operates through the orchestration of events, where indirect control enables creativity to engage with diverse media and effects, providing a conceptual framework for design that is adaptive to the cyber-physical spaces of IR4. While most chapters are devoted to computational or robotic methods, the next chapter commences with an emphasis on material engagement, where creativity operates within spatiotemporal physical events. In Chapter Three, physical acts of fabrication or assembly are leveraged within the creative design process to expand material expression, while critically, they also offer a reduction in production-oriented material waste, impacting both design outcomes and production processes.

Notes

1. *While these conditions might be resolvable by revising immigration or government spending policies, for various reasons such approaches are not always deemed politically or economically feasible.*

2. *Shimizu™, Takenaka Corp™, Tokyu Construction Co™, Komatsu™, Hazana Ando Corp™ and others.*

3. *Also referred to as "articulated robots", industrial robot arms are made by companies ABB™, Kuka™, Universal Robots™, Staubli™, Fanuc™ and others.*

Image Credits

Figures 2.1-2.2. *Robert Stuart-Smith.*

References

Aejmelaeus-Lindström, Petrus, Jan Willmann, Skylar Tibbits, Fabio Gramazio, and Matthias Kohler. 2016. "Jammed Architectural Structures: Towards Large-Scale Reversible Construction." Granular Matter 18 (2): 28. doi:10.1007/s10035-016-0628-y.

Anton, Ana, P Bedarf, A Yoo, L Reiter, et al. 2020. "Concrete Choreography: Prefabrication of 3D Printed Columns." In Fabricate Making Resilient Architecture.

Architecture Machine Group MIT. 1970. "Gerbils in a Computerized Environment." In Software: Information Technology: Its Meaning For Art, edited by Jack Burnham. New York: Jewish Museum.

Augugliaro, Federico, Ammar Mirjan, Fabio Gramazio, Matthias Kohler, and Raffaello D'Andrea. 2013. "Building Tensile Structures with Flying Machines." In Intelligent Robots and Systems (IROS), 2013 IEEE/RSJ International Conference On, 3487–92.

Augugliaro, Frederico, Sergei Lupashin, Michael Hamer, Cason Male, Markus Hehn, Mark W Mueller, Jan Sebastian Willmann, Fabio Gramazio, Matthias Kohler, and Raffaello D'Andrea. 2014. "The Flight Assembled Architecture Installation: Cooperative Contruction with Flying Machines." IEEE Control Systems 34 (4): 46–64. doi:10.1109/MCS.2014.2320359.

Autovol. 2022. "Autovol." https://autovol.com/.

Barandy, Kat. 2018. "'Form of Wander' Bifurcates and Weaves Over a Florida Waterfront Pier." DesignBoom, November 14. https://www.designboom.com/architecture/marc-fornes-theverymany-form-wander-tampa-florida-11-14-18/.

Bechthold, M, A Kane, and N King. 2015. Ceramic Material Systems: In Architecture and Interior Design. Birkhäuser.

Bergman, D. 2013. Sustainable Design: A Critical Guide. Architecture Briefs. New York: Princeton Architectural Press.

Betts, Mike, Graham Robinson, Charles Burton, Jeremy Leonard, Amit Sharda, and Toby Whittington. 2015. "Global Construction 2030: A Global Forecast for the Construction Industry to 2030." www.globalconstruction2030.com.

Bhooshan, Shajay, Tom van Mele, and Philippe Block. 2020. "Morph & Slerp." doi:10.1145/3424630.3425413.

Blueprint Robotics. 2023. "A Better Way to Build." Accessed July 25. https://www.blueprint-robotics.com/.

Bock, T, and T Linner. 2016a. Construction Robots: Volume 3: Elementary Technologies and Single-Task Construction Robots. Cambridge University Press.

———. 2016b. Site Automation. Cambridge Handbooks on Construction Robotics. Cambridge University Press.

Braithwaite, Adam, Talib Alhinai, Maximilian Haas-Heger, Edward McFarlane, and Mirko Kovač. 2018. "Tensile Web Construction and Perching with Nano Aerial Vehicles." In Robotics Research, 71–88. Springer.

Brayer, Marie-Ange. 2015. "Frank Gehry: The Interlacing of the Material and the Digital." In Frank Gehry, edited by A Lemonier and F Migayrou. Prestel.

Brooks, Rodney A (MIT). 1985. "A Robust Layered Control System for a Mobile Robot. AI Memo 864." ieeexplore.ieee.org, September.

Buchanan, C, and L Gardner. 2019. "Metal 3D Printing in Construction: A Review of Methods, Research, Applications, Opportunities and Challenges." Engineering Structures. doi:10.1016/j.engstruct.2018.11.045.

Bureau of Labor Statistics. 2021. "National Census of Fatal Occupational Injuries in 2020."

Buswell, R A, R C Soar, A G F Gibb, and A Thorpe. 2007. "Freeform Construction: Mega-Scale Rapid Manufacturing for Construction." Automation in Construction 16 (2). doi:10.1016/j.autcon.2006.05.002.

Castillo y López, Jose Luis Garciá del. 2019. "Robot Ex Machina A Framework for Real-Time Robot Programming and Control." In Ubiquity and Autonomy - Paper Proceedings of the 39th Annual Conference of the Association for Computer Aided Design in Architecture, ACADIA 2019.

Clifford, Brandon, Wes McGee, and Mackenzie Muhonen. 2018. "Recovering Cannibalism in Architecture with a Return to Cyclopean Masonry." Nexus Network Journal 20 (3): 583–604. doi:10.1007/s00004-018-0392-x.

Daas, M, and A J Wit. 2018. Towards a Robotic Architecture. ORO Editions.

Day, N, A Brown-May, and C Coney. 2005. *Federation Square*. Hardie Grant Books.

Doerfler, Kathrin, Timothy Sandy, Markus Giftthaler, Fabio Gramazio, Matthias Kohler, and James Buchli. 2016. "Mobile Robotic Brickwork." *Robotic Fabrication in Architecture, Art and Design*. doi:10.1007/978-3-319-26378-6_15.

Doerstelmann, Moritz, Jan Knippers, Valentin Koslowski, Achim Menges, Marshall Prado, Gundula Schieber, and Lauren Vasey. 2015. "ICD/ITKE Research Pavilion 2014–15: Fibre Placement on a Pneumatic Body Based on a Water Spider Web." *Architectural Design* 85 (5). John Wiley & Sons, Ltd: 60–65. doi:https://doi.org/10.1002/ad.1955.

Dusty Robotics. 2022. "Construction Robots: BIM-Driven Layout." https://www.dustyrobotics.com/.

Etherington, Rose. 2008. "Watercube by PTW Architects." *Deezeen*. February 6. https://www.dezeen.com/2008/02/06/watercube-by-chris-bosse/.

Farmer, Mark. 2016. "The Farmer Review of the UK Construction Model: Modernise or Die." London.

FBR Ltd. 2018. "Hadrian X® | Outdoor Construction & Bricklaying Robot from FBR." https://www.fbr.com.au/view/hadrian-x.

Feringa, Jelle, Wes McGee, and Asbjørn Søndergaard. 2013. "Processes for an Architecture of Volume." In *Rob | Arch 2012: Robotic Fabrication in Architecture, Art, and Design*, edited by Sigrid Brell-Çokcan and Johannes Braumann, 62–71. Vienna: Springer Vienna. doi:10.1007/978-3-7091-1465-0_5.

Fuller, Thomas. 2020. "Why Does It Cost $750,000 to Build Affordable Housing in San Francisco?" *The New York Times*. February 20. https://www.nytimes.com/2020/02/20/us/California-housing-costs.html.

Galloway, Kevin C, Rekha Jois, and Mark Yim. 2010. "Factory Floor: A Robotically Reconfigurable Construction Platform." In *Robotics and Automation (ICRA), 2010 IEEE International Conference On*, 2467–72.

GCR Staff. 2021. "Robot Digger Works 24 Hours on Its Own in Auto-Machinery Breakthrough." *Global Construction Review*. July 13. https://www.globalconstructionreview.com/robot-digger-works-24-hours-its-own-auto-machinery/.

Gerrard, Neil. 2019. "Mace Claims Evolved 'Jump Factory' Will Boost Productivity Sixfold - Construction Management." *Construction Management*. December 5. https://constructionmanagement.co.uk/mace-claims-evolved-jump-factory-will-boost-produc/.

Gramazio, F, M Kohler, and J Willmann. 2014. *The Robotic Touch: How Robots Change Architecture*. Park Books.

Gramazio, Fabio, Matthias Kohler, and Jan Willmann. 2014. "Authoring Robotic Processes." *Architectural Design* 84 (3). John Wiley & Sons, Ltd: 14–21. doi:https://doi.org/10.1002/ad.1751.

Grieves, Michael, and John Vickers. 2016. "Digital Twin: Mitigating Unpredictable, Undesirable Emergent Behavior in Complex Systems." In *Transdisciplinary Perspectives on Complex Systems: New Findings and Approaches*. doi:10.1007/978-3-319-38756-7_4.

Hahm, Soomeen. 2019. "Augmented Craftsmanship Creating Architectural Design and Construction Workflow by Augmenting Human Designers and Builders." In *Ubiquity and Autonomy - Paper Proceedings of the 39th Annual Conference of the Association for Computer Aided Design in Architecture, ACADIA 2019*.

HM Government. 2013. "Construction 2025: Industrial Strategy: Government and Industry in PartnershipConstruction 2025: Industrial Strategy: Government and Industry in Partnership." London.

HSE. 2021. "Workplace Fatal Injuries in Great Britain, 2021." London.

iNEWS. 2022. "Look!Robot Builds a House." *INEWS*. February 24. https://inf.news/en/tech/a74d5dbb7b8b0dacb8866e9e84401590.html.

Intelligent City. 2023. "Adaptive Mass Timber Automation Becomes a Reality at Intelligent City." Accessed October 29. https://intelligent-city.com/announcements/adaptive-automation-becomes-a-reality-at-intelligent-city/.

International Labour Organization. 2015. "Construction: A Hazardous Work." *International Labour Organization (ILO)*. March 23. https://www.ilo.org/global/topics/safety-and-health-at-work/areasofwork/hazardous-work/WCMS_356576/lang--en/index.htm.

Jahn, Gwyllim, Andrew John Wit, and James Pazzi. 2019. "[BENT] Holographic Handcraft in Large-Scale Steam-Bent Timber Structures." In *Ubiquity and Autonomy - Paper Proceedings of the 39th Annual Conference of the Association for Computer Aided Design in Architecture, ACADIA 2019*.

Janicki, Hubert, and Erika McEntarfer. 2015. "Where Did All the Construction Workers Go?"

US Census Bureau. October 16. https://www.census.gov/newsroom/blogs/research-matters/2015/10/where-did-all-the-construction-workers-go.html.

Jipa, Andrei, Mathias Bernhard, Mania Aghaei Meibodi, and Benjamin Dillenburger. 2016. "3D-Printed Stay-in-Place Formwork for Topologically Optimized Concrete Slabs." In TxA Emerging Design + Technology. doi:10.3929/ETHZ-B-000237082.

Johns, Ryan Luke, and Jeffrey Anderson. 2018. "Interfaces for Adaptive Assembly." In Recalibration on Imprecision and Infidelity - Proceedings of the 38th Annual Conference of the Association for Computer Aided Design in Architecture, ACADIA 2018.

Jokic, Sasha, Petr Novikov, Shihui Jin, Stuart Maggs, Cristina Nan, and Dori Sadan. 2022. "Minibuilders." IAAC. Accessed April 14. http://robots.iaac.net/.

Jones, Jenny. 2015. "Citation: Co-Robotics and Construction: OSCR 1-4 Prototypes." Architect Magazine, July 27. https://www.architectmagazine.com/awards/r-d-awards/citation-co-robotics-and-construction-oscr-1-4-prototypes_o.

Kayser, Markus, Levi Cai, Christoph Bader, Sara Falcone, Nassia Inglessis, Barrak Darweesh, João Costa, and Neri Oxman. 2019. "FIBERBOTS: Design and Digital Fabrication of Tubular Structures Using Robot Swarms." In Robotic Fabrication in Architecture, Art and Design 2018. doi:10.1007/978-3-319-92294-2_22.

Keating, Steven J., Julian C. Leland, Levi Cai, and Neri Oxman. 2017. "Toward Site-Specific and Self-Sufficient Robotic Fabrication on Architectural Scales." Science Robotics 2 (5): 15.

King, Nathan, and Jonathan Grinham. 2013. "Automating Eclipsis." In Rob | Arch 2012, edited by Sigrid Brell-Çokcan and Johannes Braumann, 214–21. Vienna: Springer Vienna.

Leadenhall Building. 2023. "Construction - The Leadenhall Building." The Leadenhall Building Development Company. https://theleadenhallbuilding.com/the-building/our-vision/construction/.

Maier, Juergen. 2017. "Made Smarter: Review 2017." London.

Mairs, Jessica. 2016. "AA Design & Make Students Build Woodland Barn with Robot." Dezeen. February 21. https://www.dezeen.com/2016/02/23/architectural-association-students-london-robotically-fabricated-barn-dorset-woodland/.

McKinsey Global Institute. 2017. "Reinventing Construction: A Route To Higher Productivity: Executive Summary."

Meibodi, Mania Aghaei, Andrei Jipa, Rena Giesecke, Demetris Shammas, Mathias Bernhard, Matthias Leschok, Konrad Graser, and Benjamin Dillenburger. 2018. "Smart Slab: Computational Design and Digital Fabrication of a Lightweight Concrete Slab." In Recalibration on Imprecision and Infidelity - Proceedings of the 38th Annual Conference of the Association for Computer Aided Design in Architecture, ACADIA 2018.

Menges, Achim. 2015a. "Fusing the Computational and the Physical: Towards a Novel Material Culture." Architectural Design 85 (5). John Wiley & Sons, Ltd: 8–15. doi:https://doi.org/10.1002/ad.1947.

———. 2015b. "The New Cyber-Physical Making in Architecture: Computational Construction." Architectural Design 85 (5). John Wiley & Sons, Ltd: 28–33. doi:https://doi.org/10.1002/ad.1950.

Menges, Achim, and Jan Knippers. 2011. "ICD/ITKE Research Pavilion 2010," July, 10–15.

Minsky, Marvin. 2011. "Minsky Arm 1967–1973." In The MIT 150 Exhibition, edited by Deborah Douglas. Cambridge: MIT Museum. https://mitmuseum.mit.edu/collections/object/2000.008.001.

Mohamed, Hesameddin, Ding Wen Bao, and Roland Snooks. 2021. Super Composite: Carbon Fibre Infused 3D Printed Tectonics. doi:10.1007/978-981-33-4400-6_28.

MX3D. 2020. "MX3D Bridge." MX3D.

Nicholas, Paul, Esben Clausen Norgaard, Mateusz Zwierzycki, Scott Leinweber, Riccardo La Magna, and Christoph Hutchinson. 2020. "A Bridge Too Far." In CITA Complex Modelling, edited by Mette Ramsgaard Thomsen, Martin Tamke, Paul Nicholas, and Phil Ayres, 260–301. Riverside Architectural Press.

NSF. 2023. "Cyber-Physical Systems." National Science Foundation. https://www.nsf.gov/news/special_reports/cyber-physical/.

Osmani, Mohamed. 2011. "Construction Waste." In Waste, 207–18. Elsevier Inc. doi:10.1016/B978-0-12-381475-3.10015-4.

Petersen, Kirstin H, Nils Napp, Robert Stuart-Smith, Daniela Rus, and Mirko Kovac. 2019. "A Review of Collective Robotic Construction." Science Robotics 4 (28). doi:10.1126/scirobotics.aau8479.

Petersen, Kirstin, Radhika Nagpal, and Justin Werfel. 2011. "TERMES: An Autonomous Robotic System for Three-Dimensional Collective Construction." Robotics: Science and Systems Conference VII.

Pigram, Dave, and Wes McGee. 2011. "Formation Embedded Design: A Methodology for the Integration of Fabrication Constraints into Architectural Design." In Integration Through Computation - Proceedings of the 31st Annual Conference of the Association for Computer Aided Design in Architecture, ACADIA 2011. doi:10.52842/conf.acadia.2011.122.

Prasath Kumar, V. R., M. Balasubramanian, and S. Jagadish Raj. 2016. "Robotics in Construction Industry." Indian Journal of Science and Technology 9 (23). doi:10.17485/ijst/2016/v9i23/95974.

Rael, Ronald, and Virginia San Fratello. 2017. "Clay Bodies: Crafting the Future with 3D Printing." Architectural Design 87 (6). Conde Nast Publications, Inc.: 92–97. doi:10.1002/ad.2243.

Reynolds, C. W. 1999. "Steering Behaviors for Autonomous Characters." Game Developers Conference, 763–82. doi:10.1016/S0140-6736(07)61755-3.

Rogers, David. 2021. "'Landmark Event' Claimed as Robots Pave Busy Highway in China." Global Construction Review. October 11. https://www.globalconstructionreview.com/landmark-event-claimed-as-robots-pave-busy-highway-in-china/.

Rust, Romana, David Jenny, Fabio Gramazio, Mat-Thias Kohler, R Rust, D Jenny, F Gramazio, and M Kohler. 2016. "SPATIAL WIRE CUTTING Cooperative Robotic Cutting of Non-Ruled Surface Geometries for Bespoke Building Components," In CAADRIA 2016 - Living Systems and Micro-Utopias: Towards Continuous Designing. 529–38. doi:10.52842/conf.caadria.2016.529.

Sabin, Jenny E. 2010. "Digital Ceramics: Crafts-Based Media for Novel Material Expression & Information Mediation at the Architectural Scale." In Life In:Formation: On Responsive Information and Variations in Architecture - Proceedings of the 30th Annual Conference of the Association for Computer Aided Design in Architecture, ACADIA 2010.

Schwab, K. 2017. The Fourth Industrial Revolution. New York: Crown Publishing Group.

Science and Technology Select Committee. 2018. "Off-Site Manufacture for Construction: Building for Change." London.

Seo, Jungwon, Mark Yim, and Vijay Kumar. 2013. "Assembly Planning for Planar Structures of a Brick Wall Pattern with Rectangular Modular Robots," 1016–21.

Shafto, Mike, M Conroy, R Doyle, and E Glaessgen. 2010. "NASA DRAFT Modeling, Simulation, Information Technology & Processing Roadmap: Technology Area 11." NASA Technology Area. Washington D.C.

Soler, Vicente, Gilles Retsin, and Manuel Jimenez Garcia. 2017. "A Generalized Approach to Non-Layered Fused Filament Fabrication." In Disciplines and Disruption - Proceedings Catalog of the 37th Annual Conference of the Association for Computer Aided Design in Architecture, ACADIA 2017.

Spong, Mark W, Seth Hutchinson, and M Vidyasagar. 2020. Robot Modeling and Control. Newark, UNITED KINGDOM: John Wiley & Sons, Incorporated. http://ebookcentral.proquest.com/lib/upenn-ebooks/detail.action?docID=6038330.

Stuart-Smith, Robert. 2016. "Behavioural Production: Autonomous Swarm-Constructed Architecture." Architectural Design 86 (2). John Wiley & Sons, Ltd: 54–59. doi:https://doi.org/10.1002/ad.2024.

Tibbits, Skylar. 2017. "From Automated to Autonomous Assembly." Architectural Design 87 (4). John Wiley & Sons, Ltd: 6–15. doi:https://doi.org/10.1002/ad.2189.

Touhey, Max. 2016. "SHoP Tops out World's Tallest Modular Building in Brooklyn." DesignBoom. https://www.designboom.com/architecture/shop-architects-461-dean-street-pacific-park-brooklyn-new-york-05-15-2016/.

Trimble. 2022. "Spot® — Trimble's Robotic Autonomous Scanning Solution." https://fieldtech.trimble.com/en/product/spot.

UKRI. 2021. "Transforming Construction Challenge – UKRI." https://www.ukri.org/our-work/our-main-funds/industrial-strategy-challenge-fund/clean-growth/transforming-construction-challenge/.

UN DESA. 2014. "World's Population Increasingly Urban with More than Half Living in Urban Areas." United Nations Department of Economic and Social Affairs. July 10. https://www.un.org/en/development/desa/news/population/world-urbanization-prospects-2014.html.

van Mele, Tom, Benjamin Dillenburger, Michael Hansmeyer, Matthias Kohler, Matthias Rippmann, Timothy P Wangler, Ena Lloret, et al. 2019. "KnitCrete: Stay-in-Place Knitted Fabric

Formwork for Complex Concrete Structures." In Cement and Concrete Research. Vol. 1.

Webb, Michael. 2008. "National Stadium in Beijing by Herzog and de Meuron." The Architectural Review. July. https://www.architectural-review.com/today/national-stadium-in-beijing-by-herzog-and-de-meuron.

Weiss, S I. 2013. Product and Systems Development: A Value Approach. Wiley.
Wilson, Wendy, and Cassie Barton. 2023. "Tackling the Under-Supply of Housing in England." London.

Wu, Hao, Ming Lu, XinJie Zhou, and Philip F. Yuan. 2022. "Application of 6-Dof Robot Motion Planning in Fabrication." In Proceedings of the 2021 DigitalFUTURES. doi:10.1007/978-981-16-5983-6_31.

Yablonina, M, M Prado, E Baharlou, T Schwinn, and A Menges. 2017. "Mobile Robotic Fabrication System for Filament Structures, in Fabricate – Rethinking Design and Construction." In Fabricate Conference 2017, edited by Achim Menges, Bob Sheil, Ruairi Glynn, and Marilena Skavara, 202–209. Stuttgart: UCL Press.

Yuan, P F, N Leach, and A Menges. 2018. Digital Fabrication. Tongji University Press.

Zhang, Ketao, Pisak Chermprayong, Feng Xiao, Dimos Tzoumanikas, Barrie Dams, Sebastian Kay, Basaran Bahadir Kocer, Alec Burns, Lachlan Orr, Talib Alhinai, Christopher Choi, Durgesh Dattatray Darekar, Wenbin Li, Steven Hirschmann, Valentina Soana, Shamsiah Awang Ngah, Clément Grillot, Sina Sareh, Ashutosh Choubey, Laura Margheri, Vijay M Pawar, Richard J Ball, Chris Williams, Paul Shepherd, Stefan Leutenegger, Robert Stuart-Smith & Mirko Kovac. 2022. "Aerial Additive Manufacturing with Multiple Autonomous Robots." Nature 609 (7928): 709–17. doi:10.1038/s41586-022-04988-4.

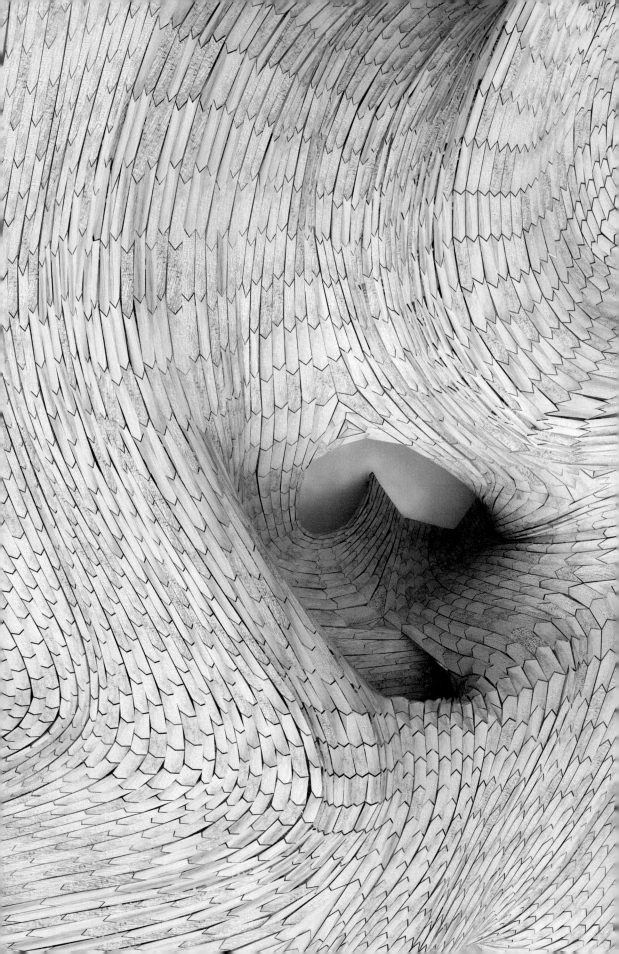

◀◀
3.1 Helsinki Library
*Close-up of ceiling
surface and atrium viewed
from above. The design
emerges from a generative
assembly process.*

3 | Generative Production
The Agency of Things

Material Engagement and The Agency of Things

Despite an intimate historical relationship, present-day building and design activities are distinctly separate operations. However, with emerging autonomous manufacturing and design technologies, more meaningful dialogue could be developed between the two. While it is uncommon for architects, fabricators, or construction companies to collaborate on creative approaches to fabrication or construction, there are exciting opportunities to directly engage with the physical world in a more creative manner. Whether one recognises it or not; materials, objects, and forces operate on buildings and help determine the nature of their existence. Conscious engagement with these expands creative freedoms and provides a means to re-think the way we produce architecture.

Within Oxford University's School of Archaeology, Dr Lambros Malafouris has been developing a theory of Material Engagement, where things and actions are viewed as cognitive extensions of the human body. Malafouris argues that humans are not the only ones capable of initiating causal events, suggesting that inanimate objects partake in such activities also. For Malafouris, causal agency is determined through the *"interface between bodies, brains and things"* (Malafouris and Renfrew 2013, 6). In his paper *At the Potter's Wheel: An Argument for Material Agency*, Malafouris suggests that a human's sense of agency in initiating creative acts is partially an illusion (Malafouris 2008). To illustrate his point, Malafouris turns to the design and manufacturing of ceramics. Using the example of the ceramist's work on the potter's wheel to sculpt clay, he portrays three parties involved in the task; the human potter, the wheel and wet clay. While the general assumption is that the potter operates through conscious predetermined intentions to rotate the wheel and to shape the clay with their hands, Malafouris describes how the wheel acts like a *"dynamic attractor"*, influencing not only the motion of the clay, but also by evoking a feedback response from the potter's hands, and in turn, affecting the potter's subsequent decisions (Malafouris 2008, 28). Malafouris suggests that much of the potter's actions precede a conscious will to act, with the brain essentially re-writing history as it comes up to speed with the state of actions in play. To support this, he cites research by Benjamin Libet that demonstrated a conscious will-to-act operates 350 milliseconds or more behind a human's 'readiness potential' and 'motor action' (Libet 1999; Malafouris 2008, 27).

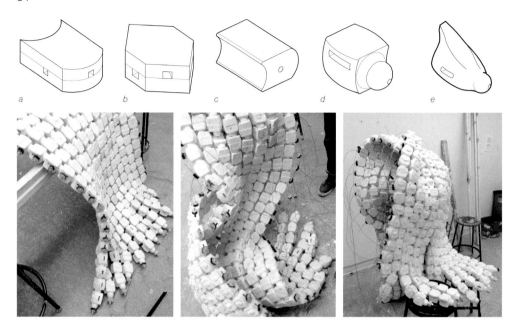

a   b   c   d   e

▲

3.3 materialAgencies
*Half-scale plaster
prototypical assemblies
using element from
Figure 3.2d above.
Different degrees of
connectivity produce
different assembly
effects such as tassels,
curvature inflections,
and self-supporting
arching conditions.*

For Malafouris, agency does not lie solely with the potter, it is shared amongst multiple actors, both human and non-human (such as the wheel). Actors, materials and actions operate in a dynamic interplay with feedback, where previous acts cause effects that constrain and influence future acts. Agency is a temporal, emergent effect of activity (Malafouris 2008, 35). Bruno Latour would call these agencies *"actants"*, sources of action that can be human or non-human, capable of producing effects and altering the course of events (Latour 2009, 237). For the political theorist Jane Bennett, Latour's actants don't operate alone, their agency arises through *"collaboration, cooperation or interaction with other bodies and forces"* (Bennett 2010, 231). Malafouris, Latour and Bennett's ideas on agency appear to align with the Speculative Realist philosopher, Levi Bryant's concept of a *"flat ontology"* where a human is considered an object, able to influence and be influenced to the same degree as other things (Bryant 2011). While Bryant does not argue that all things are equally influential, his flat ontology articulates a Posthuman position on human agency – where the human is one of many actors in a universe of object-oriented interactions. To the unacquainted, Posthumanism can sound anti-human, but this is not the case. Instead, as the contemporary philosopher Rosi Braidotti argues, Posthumanism represents a shift away from Enlightenment era Humanism that perceived the white-European male as the exemplar of rational thought to the exclusion of other genders, races, peoples, species, etc, to a more inclusive and ethical re-casting of the individual's identity within a larger cosmos, shifting humanity from paternal caretaker to an equal collaborator with the environment, and other species (Braidotti 2013, 65).

While ethically and theoretically compelling, how does Malafouris' concept of *"Material Engagement"* and Bryant's *"Flat Ontology"* relate to the way we conceive or make buildings? There are numerous contributors to building design; from human collaborators in a wide range of disciplines through to software or physical site territories that embody constraints

3.2 materialAgencies
Different prototypes of
an individual element.
Each was tested in
assembled prototypes.
The geometry of each
profile directly impacted
assembly possibilities.
a-c) initial unsuccessful
prototypes were too
rigid and could not
accommodate variable
curvature easily. d)
simple preliminary
version of final geometry,
e) final geometry.

and opportunities. There are also more remote contributors such as economic markets or climate. Beyond these conventional aspects, however, buildings embody diverse materials in solid, liquid and gaseous form that are manufactured and assembled by dynamic physical forces that are able to be engaged with creatively – extending Malafouris' potter's wheel principle to the scale of a building, or prefabricated building element. To explore this, materials and fabrication activities must be recast as *co-actants*, as essential design participants. Physical prototyping then becomes a forum, where materials and forces influence the creative process, with their behaviour informing the designer/s, impacting their intent and ultimately, design outcomes.

The projects presented in this chapter each embody a design approach that operates through direct engagement with materials, fabrication, and assembly processes. They required hands-on experience with physical world processes: such as material viscosity or dynamic physical forces that operate under gravity. To develop design proposals, creative work was iterative, and involved continuous adjustment of design-intent, incorporating feedback from material prototypes. In most of these works, any graphical representation of the design or manufacturing tasks played a very minor role during design development. *Design operated primarily through the orchestration of events, strategised through prototypes and event-based diagrams.* In these creative acts, objects and materials have agency, particularly within generative approaches that leverage material dynamics or assembly.

3.4 materialAgencies
Partial assembly mock-up
of final precast concrete
elements using geometry
prototype from Figure 3.2e.

Generative Assembly

At Stuttgart's Institute for Lightweight Structures in the 1960s, Frei Otto developed an approach to design that leveraged physical forces operating within materials to derive efficiently designed forms. Natural processes of self-organisation were studied, and re-appropriated

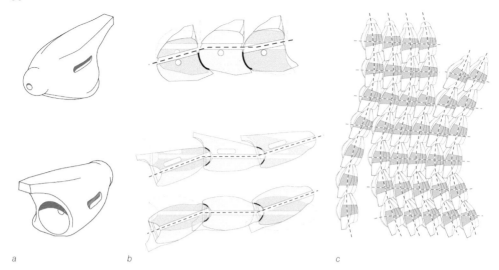

a   b   c

▲

3.5 materialAgencies
Individual element
geometric profiling and
connection details were
designed to impact
part-to-part alignments
in larger assemblies: a)
two views of the final
element geometry, b) joint
profiling through both axis
of cables, c) assemblage
with elements threaded
through two cable axes of
cable. Slotted connections
are shown in grey.

as form-finding methods for building design. Otto studied soap-film's natural tendency to minimise surface tension and harnessed this physical phenomenon to develop minimal-surface tensile structural forms for canopy roofs such as the Munich Olympic Stadium (Schanz 1995). He also developed an ultra-thin timber grid-shell structural system that could be manufactured as a simple, regular grid and adapted into formally complex large-span buildings such as the Multihalle in Manheim (Schanz 1995, 140). Otto's grid-shells and tensile structures took their final shape through the act of assembly on the construction site, where their precise positioning caused a negotiation between designed matter and physical forces.

Otto initially developed his designs through small-scale physical prototypes. These prototypes often didn't use the intended building materials, but materials sufficiently representative of the primary physical forces that would operate on the designs at building scale. Tensile fabric structural models were developed using either soap film (preserved by freezing), or stretched spandex, together with a shaped wire profile. Grid-shell designs were conceptualised through suspended wire-mesh models that created ideal tensile dome forms under gravity that could be inverted to operate as efficient compressive structures (inspired by Hooke's funicular structural ideas (Hooke 1676) discussed in Chapter Six).

At the time of his research, Otto's methods were challenging to simulate computationally due to limitations in computing power in the 1960s and 70s. To support his physical design developments, photogrammetry was used to translate models into data that could be used for structural analysis and building set-out (Schanz 1995). Following advances in computing over the last half-century, a whole suite of software solutions are now available that are capable of computing Otto's form-finding processes rapidly. This has given rise to an overabundance of their use in building designs. Amongst this stands some novel research such as Sean Ahlquist's research into tensile structures (Ahlquist, Erb, and Menges 2015) or Achim Menges' fibre-composite pavilions (Menges 2012) that innovate by extending Otto's principles into new territories. Beyond these research exceptions, however, the ubiquity of project-led applications of Otto's methods ad infinitum by others within the field of architecture produces a formal character that persists unchanged across a multitude of projects, one that operates at the scale of a building's entire geometry due to its emphasis on homogenising the relations between matter and force, thereby homogenising architectural design expression.

It is possible though, to produce alternative design character within processes similar to those of Frei Otto, if material and physical forces are organised heterogeneously. Depending on the designer's orchestration of matter–force relations, design results might be structurally efficient or inefficient, most, somewhere in-between. Given that architecture embodies several often-conflicting design considerations, such in-betweens are often desirable. Extending Otto's form-finding methods to heterogeneous material organisations enables material forces to self-organise within an assembly or fabrication process to produce differentiated design effects, where design character is variable, and directly related to material expression.

During a visiting professorship at Washington University in St Louis, the fabrication studio *materialAgencies*[1] explored a generative approach to assembly within a small, fabricated installation. The studio, run by the author and Robert Booth, sought to hybridise principles of cable-net suspension, post-tension compression and drapery within a singular material assemblage (Stuart-Smith 2018). While cable-net suspension and post-tension compression are relatively efficient, drapery provides moments of material excess, an inefficiency that also contributes to design expression. It was determined from the outset that heavy mass-produced precast concrete elements would be arrayed and connected together by a steel cable-net in two dimensions and tailored into a three-dimensional arrangement so that the material assemblage had opportunities to

▼

3.6 materialAgencies
Photograph mid-way through final assembly. The majority of elements were assembled flat on the ground and lifted into their final formation with the scaffold gantry.

3.7 materialAgencies

Close-up showing the overlap and variable orientation of assembled parts. Directional grain is generated through tail-like extensions that exaggerate the overlapping of elements in one axis.

operate in compression when tightly packed, or collapse in moments of suspension (tension). In contrast to self-assembly research by Skylar Tibbits and others that involves either agitation or submersion within fluid to bring objects together (Tibbits 2016), *materialAgencies* focused on more explicitly defined object-to-object connectivity impacting design formation and expression. The array of objects was to collide with each other and self-organise under gravity, constrained by their interconnections, in order to arrive at a final formation. Interconnections would include pleating, pinching or edge-to-edge joining to connect the arrayed objects together in a similar manner to the tailoring of a garment or curtain. Due to the physics principles at play, it was concluded that material prototypes would be essential to the design development of the proposed installation, and to curate its material expression. Students iteratively fabricated and assembled prototypes whose physical collapse under gravity operated as a generative means of assembly, and informed subsequent design iterations.

Two parallel developments were undertaken that directly impacted the design result. The first was an exercise in tailoring; the array of objects could be interconnected in a multitude of ways, and the specifics of these connections determined the topology of the final assemblage, and therefore also greatly impacted form-finding. A series of small-scale and half-scale models were developed using plaster-cast elements to rapidly design and test various tailored relationships (Figure 3.3). As the precise shape of each object-element affected the spatial packing and collisions each element negotiated with neighbouring elements, a second line of development involved the 3D design and physical prototyping of the physical elements (Figure 3.2). The physical form of the individual element had direct agency in determining opportunities and constraints in the overall assemblage and resulting aesthetic effects. To demonstrate variable design expression within an assemblage of identical objects, an aesthetic decision was taken to design the object to visually express a directional grain within the assemblage and to enable individual elements to vary in orientation. The element was designed to achieve these effects through three different physical strategies:

3.8 materialAgencies

Photographs of the final assembled installation. (L) Overall view (R) Close-up showing variation in orientation and overlap between elements.

a

b

Generative Production

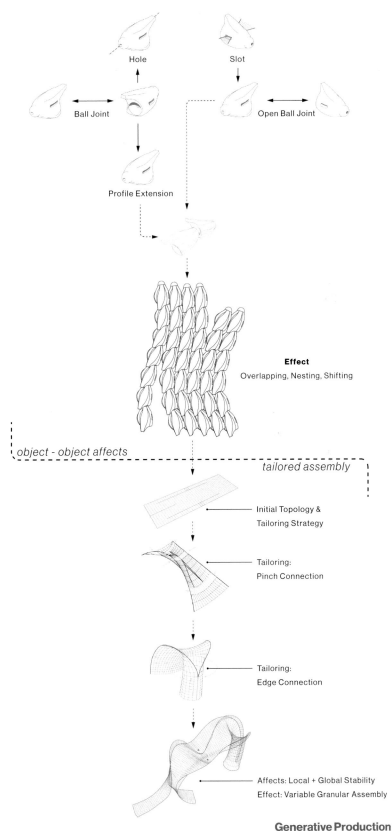

Hole

Slot

Ball Joint

Open Ball Joint

Profile Extension

Effect
Overlapping, Nesting, Shifting

object - object affects

tailored assembly

Initial Topology &
Tailoring Strategy

Tailoring:
Pinch Connection

Tailoring:
Edge Connection

Affects: Local + Global Stability

Effect: Variable Granular Assembly

Generative Production

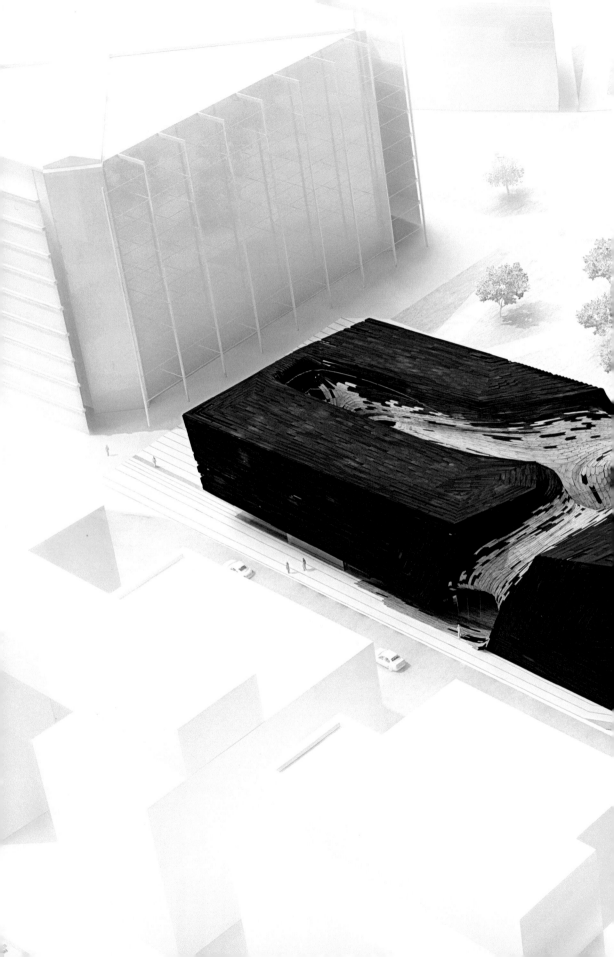

◀◀
3.11 Helsinki Library
*Aerial perspective view
of the design proposal.
The building's massing
aligns with the street
and adjacent buildings
on the left before
turning to mirror Steven
Holl's Chiasma Gallery
(partially in view top
left), and to face the
parliament building (out
of view across the park
at the top).*

1. Element-to-steel cable relations:
A 2D array of elements was connected by two axes of parallel-distributed steel cables. Each element is threaded through one cable in each axis. The assemblage of elements cannot turn around a bend if the connections of each element remain fixed at regular positions. To illustrate this, imagine a steel cable bent into a semi-circle with additional cables offset as curves on the outside. If identical objects were distributed along each cable, the outer cables would accommodate more objects. The objects on each cable would align at one end but deviate in spacing towards the other end, corresponding to the increasing length of each offset cable. To bend this arrangement into a different shape, the objects on one cable would also need to shift relative to objects on an adjacent cable. To address this, each element's two cable connections were designed differently (Figure 3.5). Elements are threaded over one cable using a hole-type connection that centres each element on the cable. As elements follow the curvature of this cable, they are designed to express this axis as a primary directional grain. The second axis cable is threaded through a slotted hole in each element, allowing the element to shift in position and orientation along this axis-2 cable, enabling various bending configurations in the assembly.

2. Element-to-element interactions:
The physical profiling of the individual element impacted how it could be positioned adjacent to other elements. Each element had two ball-like joints to connect to neighbouring elements at varying angles. Each steel cable axis had one joint, with a male and female joint profile located at opposite sides of the element. The axis-1 element-chain had a tightly enclosed ball joint for the hole-type cable connection, while the slotted-hole cable axis connection had an open, partial ball joint that allowed elements to slide. This allowed an axis-1 element-chain to shift in position relative to an adjacent one. The profile of the open ball joint was also curved to force variations in the orientation of each element relative to the specific point in which the element made contact with an adjacent element (Figure 3.5).

3. Element profiling to produce visual effects.

3.12 Helsinki Library
Interior perspective. Tight arched regions in the ceiling are self-supporting while more undulating and flat regions operate as a suspended ceiling.

A wing-like extension was added to one end of the element to visually imply directionality. The ball-joint profiling created subtle shifts in each part's orientation that were visually amplified by the wings. The wings also exaggerated variations in visual overlap between elements and variations in curvature within the assemblage – by projecting the element's profile further beyond the line of the assembly in areas with tighter bends (Figures 3.4, 3.7).

The final installation design was developed through experimentation and testing of half-scale prototypes in plaster (Figure 3.3), and small assemblies using full-scale precast concrete elements (Figure 3.4). Insights gained from these prototypes were used to create an abstract tailoring template (Figure 3.10) that outlined the sequence of assembly operations necessary to realise the final installation (Figures 3.8, 3.9). This tailoring template served as the sole construction drawing, defining connections between elements within the array. The *materialAgencies* installation's design was arrived at through an act of assembly – a generative design process where design agency operated within a series of planned events that determined the interconnections and collisions of object-elements (Figure 3.10). To a certain extent, these elements also had some degree of agency. Similar to Malafouris' description of the potter's wheel operating as a strange attractor that influenced the potter's actions, in *materialAgencies*, iterative prototyping enabled object-to-object interactions to inform subsequent iterations of design development.

3.13 Helsinki Library
Perspective view of exterior near park-side entry. Warped creases and shifting alignments in timber elements above the entry are characteristic features from the tailored generative assembly process, caused by interconnections between elements. These local effects provide a bespoke character to the project that is inseparable from its production method.

The design principles of the *materialAgencies* installation were also explored at an architectural scale in Robert Stuart-Smith Design's proposal for the 2012 *Helsinki Library* Design Competition in Finland (Glancey and Léglise 2013) (Figure 3.11). Within the design, a lightly coloured timber surface mediates between several interior and exterior spaces, creating variable threshold conditions (Figures 3.14, 3.15).

48

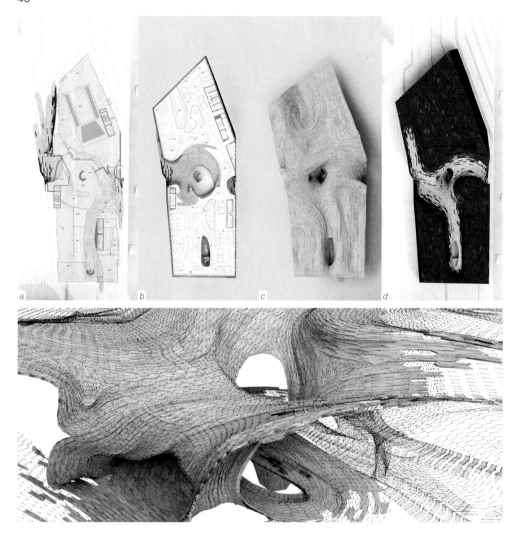

a   b   c   d

◀◀
3.14 Helsinki Library
A continuous timber
assemblage operates as
a polyvalent architectural
element that mediates
the exterior and interior,
combining entry steps/
amphitheatre, circulation,
atrium, false ceiling
and self-supporting
atrium-column surface.
During these shifts in
use, the one material
system transitions from
self-supporting post-
tension compression to
a cable-net suspension
false ceiling.

The surface, situated within an otherwise regular laminated timber-structured building, operates as atrium, ceiling, roof, circulation space and, as a park-side entry stair (Figures 3.11, 3.13). Given the scale of the building proposal and its conceptual stage, design via prototyping was not an option, nor was detailing of the timber elements necessary. The competition entry's role was to communicate the intentions of the design in the hope that it would be shortlisted for continued development (it was awarded 3rd place in a public exhibition and vote from over 400 entries; however, unfortunately, it did not attract the same vote from the competition jury). In this context, the design was developed partly through a custom-developed algorithm and partly by explicit 3D modelling.

The algorithm simulated tailoring and gravitational effects on an assembly of the lighter-coloured timber elements (an adaptation of the software *bodyTopologic,* discussed in the next chapter). A matrix of discrete timber elements was abstracted to a mesh surface particle-spring network and subjected to gravitational force with fixed edge constraints, creating a

◀

3.15 Helsinki Library
*Planimetric views; a)
ground floor, b) first
floor, c) top floor ceiling,
d) roof. The public
realm operates within a
distinctive spatial and
material field defined
by a series of spatial
thresholds.*

degree of surface tension. Running as a time-based simulation, surface mesh nodes would calculate collisions and settle under gravity to a near-equilibrium condition. Parts came to rest in a similar manner to the *materialAgencies* physical prototype, approximating the formation and character of element-to-element interactions. Similar to the tailoring template of *materialAgencies*, the mesh surface operated as a 3D spatial template that was strategically shaped to connect to various edges of the overall building geometry and to the ground at support locations. Linkages across axis-1 cables were also defined that represented element-to-element physical connections. In the simulation, the surface operated as a mesh network of particles with a mass and collision radius and springs that maintained a regular length between particles (for a more detailed description of the *bodyTopologic* algorithm see Chapter Four). Linkages between vertexes were treated as infinitely strong springs, each with a rest length of zero that effectively pulled pairs of particles together[2], manipulating the surface into a result affected by both tailoring operations and gravity. The resulting surface was then articulated to embody the array of individual elements (with more time, the simulation could have been upgraded to operate as a rigid-body simulation, where collisions would be calculated using the 3D element geometries rather than using an approximated spherical collision radius at the corners of each object).

◀

3.16 Helsinki Library
*A partial 3D chunk of
the assembled surface
which operates as an
atrium, circulation route
and ceiling. The design
arises from the physical
interactions of timber
elements assembled
in a similar manner to
materialAgencies, also
expressing a vector-
based directional grain
in one axis. Due to the
conceptual design
stage of the project, the
design was developed
through a combination
of digital simulation and
explicit modelling rather
than physical assembly.*

Where *materialAgencies* produced a directional grain that was visually enhanced by wing-like extensions on the element's geometry, in the *Helsinki Library* each timber element points in one direction using a chevron profile. The chevron profile also enables the surface to appear smooth when relatively flat and stepped where the joint between chevrons opens to accommodate increasing degrees of surface curvature. Similar to *materialAgencies*, the directional grain of timber elements is further enhanced by slippage between element-chains, with alignment between elements staggered relative to variations in curvature throughout the surface. The arrow-like element geometry together with the misalignment of element-chains produces a grain-direction within the building that is aesthetically strong in reinforcing spatial and formal dynamism (Figures 3.13, 3.14).

▶

3.17 Material Order
*Photograph of Mies van
der Rohe's Barcelona
Pavilion. While material
grain is considered in
the bookmarking of
wall marble panels (a),
misalignments in the
marble wall joints to
the floor (b-c) reinforce
the design operating
as a series of planes.
Simultaneously, this
undermines the integrity
of the 3D building object
being read as a singular
entity.*

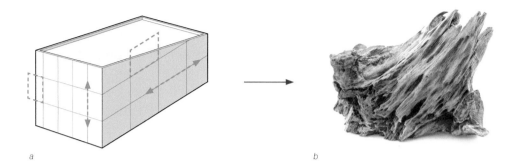

a

b

▲

3.18 Material Order
*(a) A prolific material
condition undermines
the autonomy of the
architectural object.
Cladding panelisation
can erase volumetric
integrity if:
• panel proportions
emphasise different
directional grain on each
face, emphasising each
plane,
• or if no attempt is
made to relate tiling to
building geometry or
edge conditions.
(b) In contrast, driftwood
embodies a directional
grain that is formally
related to its overall
volume and silhouette.*

Material grain is rarely expressed at the scale of buildings, except where a natural material has such strong visual patterning that alignment between adjacent panels must be considered, such as in marble cladding where grain alignment and mirroring is regularly practised (commonly referred to as "book matching") (Figure 3.17). Where grain is considered, it seldom impacts element shapes, multi-element tiling patterns, or overall building form. As most buildings are not monolithic, this disjunction between material panelisation and overall form discredits the agency of the building-object, by disabling its ability to produce a totalising affect (Figure 3.18a). In contrast to this, a piece of driftwood, expresses an anisotropic materiality through a directional grain that varies in order yet manages to relate to the object's overall figuration (Figure 3.18b). Architecture is typically devoid of driftwood-like sub-object organisational expression. The *Helsinki Library* design challenges this by organising the timber elements to express a directional alignment within the library, that results in a coherence and unity between the object-element composition of the building, its spatial organisation and volume (Figures 3.19). This unity is further enhanced by the dynamic alignment of the timber elements under gravitational force, where their organisation and tailoring direct a form-finding process that relates element-expression to surface formation, to arrive at an integrated surface character (Figure 3.12).

The *Helsinki Library* proposal's surface-assemblage operates as a hybrid system; part self-supporting atrium, part fixed floor plate, and part suspended ceiling (Figure 3.14). In the project, a digitally simulated approach to generative assembly enables the design outcome to embody spatial, formal, and physical negotiations within a variable design character that is the result of physically designed events of formation. The design is a product of vectoral and physical interactions and affects that register in the project's variable formal and material expression (Figures 3.13, 3.16). The design was developed through the orchestration of a series of events and protocols for material engagement.

Generative Fabrication

Concrete is an exciting material whose phase-changing properties embody substantial generative-design potential. During curing, concrete crystalises, transitioning from a liquid to a solid. While concrete is one of the oldest and most utilised building materials, it continues to be developed into new and innovative products through ongoing research and development. Cemex™ is one of a few international cement companies that develop their own add-mixtures in-house, enabling Cemex to exercise exceptional control over the complex properties of concrete and its distinctive characteristics within a wide range of conditions including pouring, transport, curing and use phases. Cemex's self-compacting loose-fibre reinforced *Resilia*™ concrete (Cemex 2020b) is particularly special due to its high strength-to-weight ratio and ability to remain sufficiently fluid during pouring when mixed with loose-fibre reinforcement.

A collaboration between the University of Pennsylvania's M.Arch Program 701 Design Studio Stuart-Smith[3] and Cemex™ Global R&D[4] led by Vice President Davide Zampini, explored methods of generative fabrication that exploited the phase-changing properties of concrete during two short, experimental workshops. The workshops involved classes on concrete mixtures and casting methods, and two to three days in which students developed a short design-fabrication experiment. Two of these experimental projects investigated the use of *Resilia*™ in design proposals that instrumentalised fabrication as part of the creative process. In both experimental projects, the properties and characteristics of Cemex's state-of-the-art concrete were instrumental in enabling and inspiring the fabrication approaches explored. In these experiments, design operated within a series of formative events to produce a more complex outcome than what was described in each fabrication setup. This complexity emerged from the orchestration of dynamic material processes in relation to gravitational force.

▼

3.19 Material Order
Helsinki Library object integrity: form and material organisation are integral to one-another and collectively generate a shared design expression. a) The complete building embodies a unity between material grain and its geometric volume that also integrates sub-building areas such as this partial chunk of its atrium (b).

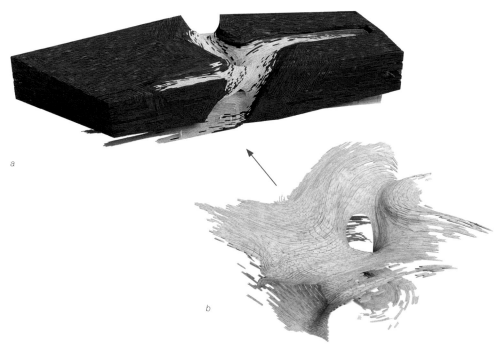

a

b

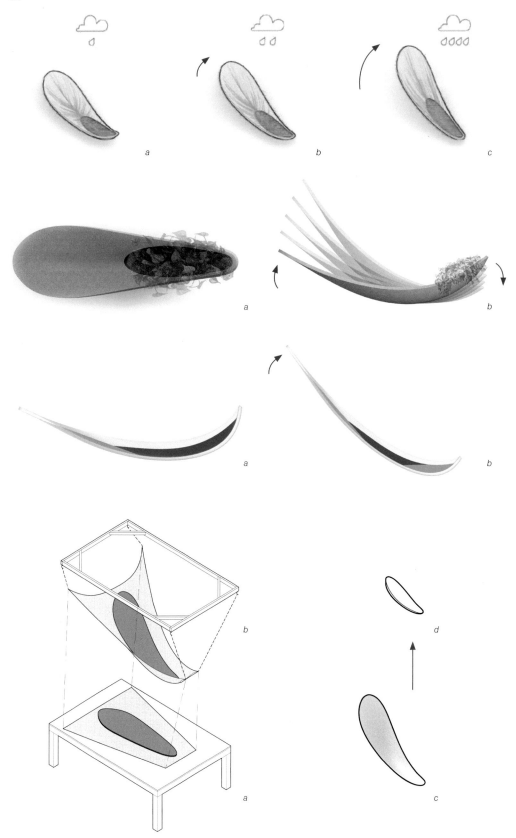

3.20 RkrPlanter
The planter is designed to collect rain-water and direct it to a soil-free plant zone that uses porous Cemex Pervia concrete as a substrate. (a-c) Rainwater collection effectiveness varies based on the tilt-angle of the planter.

3.21 RkrPlanter
a) top view of planter, b) side view of planter.

3.22 RkrPlanter
Two long sections through the planter show variations in tilt angle based on collected water's volumetric weight shifting the centre of gravity of the object. Increases in water volume reduce its capacity for water collection. The light grey material is impermiable Resilia concrete. Dark grey indicates a porous Pervia concrete that is lighter and suitable for use as a plant substrate. Blue indicates water flooding of the Pervia concrete zone.

3.23 RkrPlanter
Fabrication process: a) concrete is cast on flat geotextile fabric, b) geotextile fabric is suspended after the concrete is half-cured, shaping the cast through gravitational and material force, c) after curing the double-curved Resilia concrete part is removed from the geotextile, d) an additional porous Pervia concrete is cast into the bottom portion of (c).

Rkr

The first experimental project, *Rkr*, hybridised two existing fabrication techniques to develop an alternative approach to fabric formwork concrete casting. Fabric has been utilised as a flexible mould for fluid concrete casting in buildings designed by Miguel Fisac (Fisac and De La Sota 2013) and research conducted by Mark West (West 2016), Andrew Kudless and others (Dourtme et al. 2012; Kudless 2011) where its suppleness provides geometric design flexibilities whilst reducing material waste in formwork. More recently, Cemex together with Roger Hubeli and Julia Larsen developed research that showcased flat concrete casts being folded like origami when half-cured, avoiding volumetric formwork (Hubeli and Larsen 2016). *Rkr* hybridises these methods, casting concrete flat and forming it when half-cured by suspending it within fabric formwork to efficiently create a doubly curved concrete surface without incurring any material waste or time to produce formwork.

In the *Rkr* project, concrete is cast flat on a table, on top of a flexible geotextile fabric using only a small edge trim for a mould to define the boundary shape of the 1.5cm thick cast. After several hours, the concrete is half-cured, yet supple. Together with the fabric, the cast is raised above the table, and the fabric suspended from its four corners, causing the concrete to be shaped by the doubly curved form of the fabric, stretched under the weight of the concrete (Figures 3.23, 3.24). To design the final outcome, a particle-spring physics simulation was developed in Rhino3D's Kangaroo™ plugin (Piker 2013) to model the material behaviour of the fabric and concrete. Simulations enabled several geometric outcomes to be explored by testing the relative positions of the fabric's four suspension points. This computational approach directly informed physical set-out locations for the fabricated prototype. The physical cast was suspended in the fabric for a further ten hours, until sufficiently cured that it could be removed while retaining its double-curved shape. Typically, casting double-curved concrete elements requires the creation of volumetric formwork moulds, often produced through CNC milling, which is time-consuming and results in material waste. In this experiment, by creatively approaching the casting process, the manufacturing method was simplified to employ a reusable geotextile fabric as formwork, saving time, energy, and material. Furthermore, the design's form and material outcome were closely tied to its fabrication process and the concrete's material characteristics. The fabrication process was also only possible due to the specific engineered properties of Cemex's *Resilia*™ concrete which allowed for a degree of suppleness when the material was half-cured, supporting the manipulation of the thin cast's shape without the cast tearing or deforming.

Rkr was not only dynamically manufactured, it was also intended to be dynamic in use. The outdoor planter-pot was designed to tilt to different angles based on the amount of water it contained. When dry, it would tilt to increase its water capture surface area to collect more rainfall. When full of water, a steeper inclination angle reduced water intake, also visually indicating that manual watering was not necessary (Figures 3.20–3.22). The interior of the planter was cast with a second concrete mix – Cemex's *Pervia*™ (Cemex 2020a) which is porous and can be used as a substrate for plants instead of soil (Figure 3.24e). To determine the planter's tilt-angle range, a 3D model assessed the relative weights and centres of gravity for different ratios of the planter's two concrete mixtures. The planter was designed to utilise collected water to displace its centre of gravity,

and cause the planter to rock to a different inclination (Figure 3.22) (unfortunately the prototype was never tested with water although simulations indicated that it theoretically worked). While *Rkr* leveraged the gravitational flow of water to displace its centre of gravity, a second project, *Lava Column*, explored the gravitational flow of fluid concrete to generate a design during fabrication.

Lava Column

Apart from additive manufacturing (see Chapter Six) or limited examples of robotically actuated slip-casting (Lloret et al. 2015), concrete is typically cast in static moulds. The level of detail and complexity of these moulds directly determines the geometric results of a cast. In *materialAgencies*, elements were cast using an elaborate single-piece, all-enclosing, flexible 3D silicon mould. Creating moulds requires a significant amount of material, energy, and time. Therefore, reducing any of these aspects would be highly beneficial. When casting concrete into a complex mould, the mixture needs to have sufficient fluidity to be poured into all parts of the mould. At the same time, enabling the concrete to cure rapidly offers time-efficiency benefits. These seemingly conflicting properties can be achieved by using admixtures in the concrete mixture. For example, a super-plasticiser can improve fluidity, while an accelerator (chemical catalyst) can speed up curing.

When poured and observed outside of a mould, concrete flows in a similar way to lava. Its viscosity increases over time as it cures rapidly, eventually hardening with a rippled exterior that captures its dynamic history in solid form and adds more geometric detail. This suggests that a geometrically simple mould could be used if concrete's viscosity and phase-changing properties also contribute to defining a cast's geometry. To explore this dynamic process, a series of experiments were conducted. Different mixtures of concrete were tested, flowing down a sloped plane around simple obstacles (Figure 3.28a). These studies helped formulate event-based rules for material engagement, including the specific properties of the concrete mixture and its viscosity, together with formwork principles for the size and spacing of obstacles, and slope angles effective at distributing the material. This resulted in a design method specifically related to a particular material mixture and fabrication technique.

◀

3.24 RkrPlanter
Fabrication process:
a-b) Resilia concrete
is poured onto a flat
geotextile, c) the edge
frame is removed when
the concrete is half-
cured, d) the geotextile
with concrete is then
suspended, taking on a
double-curvature form
through gravitational
force, e) after the
Resilia cast is cured,
it is removed from the
geotextile and a layer of
Pervia concrete is cast
into its concave interior.

Following the concrete flow experiments, inspired by Antoni Gaudi's Park Guell in Barcelona (Cuito and Montes 2003), students developed a speculative design proposal for an urban park pavilion. An undulating green-roof was to be supported by a field of columns. The key feature of the design – Lava Column – was based on a concept for a column and column-arm that incorporated knowledge from the previous concrete flow experiments, adopting them as a means of material engagement that would operate creatively in the formation of the column. A single-column fabrication experiment was undertaken to test the design-fabrication principles (Figure 3.29). The column was cast upside-down, with the column-arm mould operating as an exposed slope on which a series of obstacles were arranged. Sitting on top of the tapered column-arm mould was a double-sided mould for the column which contained a thin continuous cavity, open on its underside, exposed to the column-arm mould below (Figure 3.29a). Multiple batches of a loose-fibre reinforced self-compacting concrete were poured from above into the column mould and flowed down and out onto the open column-arm mould surface. On the slope of the column-arm mould, the concrete streamed around the obstacles, creating a multi-layered lava-like branching formation, leaving voids where the obstacles were located (Figure 3.29).

Lava Column successfully demonstrated that a fluid concrete could be dynamically formed into a column-like structure using a self-compacting loose-fibre reinforced concrete mixture and a relatively simple mould, with the resulting formation embodying more geometrical complexity than the formwork explicitly defined (Figure 3.29). Less successful was the column itself; being an initial prototype, lessons were learned about the form of the column mould and its ability to facilitate material flow during the casting process. Concrete within the mould was not sufficiently distributed to allow it to be de-moulded without breakages. Blockages during pouring prevented adequate material coverage throughout the entire mould. This could easily be addressed with additional development beyond the two days of the workshop. The concrete mixture was sufficiently fluid, although it would have been beneficial for the curing rate to be slightly slower, but not so slow as to cause the concrete to continue to flow off beyond the mould before

▶

3.25 RkrPlanter
Pervia concrete is cast
into the concave interior
of the fully cured Resilia
concrete element. The
Pervia concrete is
porous and suitable for
being used as a soil-free
substrate for plants to
grow in, enabling the
combined concrete
materials to operate as
a planter.

Generative Production

▶

3.26 RkrPlanter
*Design diagram
illustrates how the
outcome and its
in-use performance
are produced in
response to a series of
design-orchestrated
events. Design
operated through the
development of this
recipe that describes
dynamic events. The
fabrication process was
enabled and inspired by
the specific concrete
mixture's properties and
ability to be manipulated
when half-cured.*

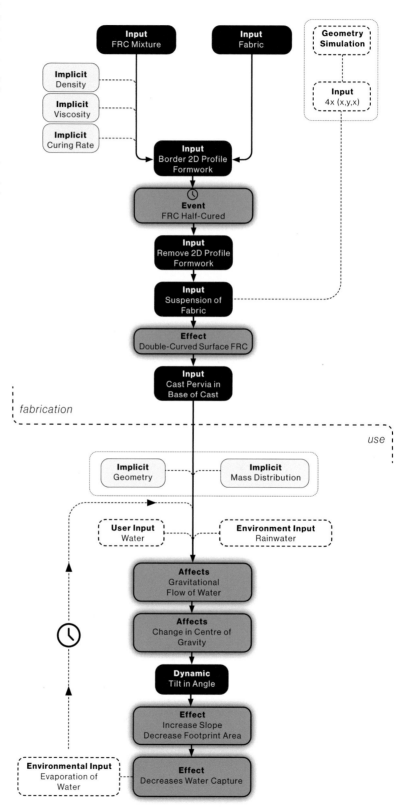

curing. Some development of the concrete mixture was undertaken during the fabrication process to fine-tune the delicate timing of the curing rate and other material properties although further development would be beneficial. Interestingly, gravity and viscous flow were observed to cause a partial alignment between many of the reinforcement fibres in the direction of gravitational flow (Figure 3.29b). Inverting the column to its correct orientation, this flow direction would roughly align with the direction of primary structural forces within the column, which could be potentially strategised in future work to provide greater tensile strength within the concrete where it is most needed.

Following the workshop, Penn's Autonomous Manufacturing Lab developed a simulation of the *Lava Column* fabrication process called *LavaSim*. In *LavaSim*, a particle-springs model simulates the concrete being poured, along with the dynamically changing properties of the concrete as it commences phase-changing whilst flowing down the formwork (such as its viscosity and friction coefficient) (Figure 3.27). The simulation also highlights a substantial spatiotemporal variation in the

▶
3.27 Lava Column
Material physics simulation study of the gravity-influenced casting process. a) Mid-simulation: red particles are cured while yellow regions remain fluid, b) rendering of a simulation result still in the mould, c) outcome inverted to its final orientation with formwork removed.

▶
3.28 Lava Column
a) Initial gravitational flow studies of concrete poured onto a slope around obstacles, b) close-up of multi-layered pouring effects captured in the simulation, c) simulation of concrete gravitational flow around entire column arm.

Generative Production

▲

3.29 Lava Column
Photographs of the final
casting experiment. The
formwork is inverted
for casting to enable
concrete to flow down
the column and over the
column-arm formwork
at its base. a-b) photo
during casting with all
formwork, c) after curing
with exterior formwork
removed. Formwork on
the inside of the column
arm is still to be removed.

state of curing within the flowing concrete, further illustrating the impact of the formwork geometry on a cascading series of material-physical events that determine a design outcome. In *Lava Column*, the design outcome was directly tied to the formwork's geometry and the material's characteristics when subjected to gravitational flow within and around the formwork. Both the experimental prototype and simulations also managed to generate a porous network of concrete for the column-arm with layered, textural aesthetics derived from the viscous flow and curing process where the concrete underwent a phase-change from liquid to solid while being subjected to gravity. A series of affects and effects were developed in this custom design-production process (Figure 3.30), demonstrating creative benefits in exploring alternative forms of material engagement within design.

Event-based Rules of Material Engagement

Projects discussed in this chapter demonstrate design activities operating within the development and testing of event-based rules of material engagement. Physical acts of fabrication or assembly were leveraged within the creative design process to produce a material-aesthetic expression, while some of the examples also offered a reduction in

▶

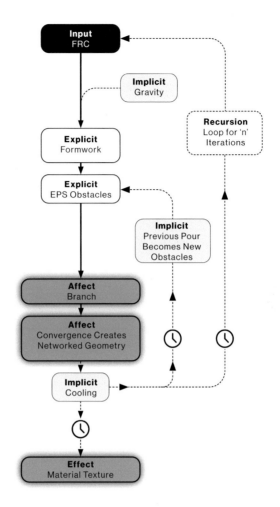

3.30 Lava Column
Design diagram illustrates how the dynamic material production process is leveraged to generate the design outcome through the orchestration of spatiotemporal events. Note in this particular design process the material properties and characteristics of the loose-fibre reinforced concrete mixture is the primary input and instrumental in determining material behaviour and the realm of possibilities for the fabrication process.

production-oriented sacrificial formwork material waste, impacting both design outcomes and production processes. *materialAgencies* and the *Helsinki Library* explored generative means of assembly (Figure 3.10), while *Rkr* and *Lava Column* involved dynamic forms of fabrication that leveraged the phase-changing properties of specific concrete mixtures as an active agent in design formation (Figures 3.26, 3.30). Similar to Malafouris' theory of Material Engagement, materials and objects (such as concrete, moulds, obstacles, fabrics, cables, cast objects, etc.) extended human design activities. These materials' physical behaviour directly contributed to design outcomes, and additionally provided feedback that influenced designers' decisions. This inclusive, participatory approach to design delegates part of the creative act to processes not directly controlled by the designer, casting fabrication as a generative design process. In the next chapter, generative design is extended into another medium – the 3D graphical space of computing. While architects frequently use computational processes for automation, optimisation or geometry rationalisation activities, an alternative semi-autonomous computational approach is presented that creatively contributes to design outcomes.

Notes

1. *The 2011 materialAgencies project was undertaken by all students together at Washington University, St Louis. Dep. Architecture Masters Level Fabrication Studio.*

2. *For an explanation on springs and computational material-physics see Chapter Six.*

3. *University of Pennsylvania, Weitzman School of Design, Department of Architecture, Masters of Architecture, final year 701 studio. Studio instructor: Robert Stuart-Smith. Studio TA: Musab Badahdah.*

4. *Cemex Global Research and Development, Brug de Bielle, Switzerland. Director: Davide Zampini. Workshop Leaders: Alexandre Guerini, Carlos Enrique Terrado, Valentina Rizzo, Matthew Meyers.*

Project Credits

materialAgencies (2012): *Washington University, St Louis. Dep. Architecture Masters Level Fabrication Studio. Instructors: Robert Stuart-Smith & Robert Booth, Students: Guru Liu, Ruogu Liu, Lu Bai, Enrique de Solo, Shuang Jiang, Christopher Thomas Quinlan, Zhe Sun, Wing Yin (Alice) Chan, Michael Chung, Chris Moy, David J Turner, Matt White.*

Helsinki Public Library (2012): *Robert Stuart-Smith Design Ltd. Project Director: Robert Stuart-Smith. Project Team: Isaie Bloch, Stella Dourtme, Felipe Escudero, Gilles Retsin. Project Consultants: Transsolar, Stuttgart.*

Rkr (2019): *University of Pennsylvania & Cemex Global R&D. University of Pennsylvania Dep. Architecture, 701 Graduate Masters level Design Studio. Instructor: Robert Stuart-Smith, TA: Musab Badahdah. Cemex Global R&D: Vice President: Davide Zampini, Workshop Leaders: Alexandre Guerini, Carlos Enrique Terrado, Valentina Rizzo, Matthew Meyers. Students: Cai Zhang, Dongyun Kim, Hao Zeng, Lingyun Yang, Mo Shen, Qi Liu, Rui Huang, Sihan Zhu, Xiaoyi Peng, Yuting Qian, Zhaoyi Liu.*

Lava Column (2018): *University of Pennsylvania & Cemex Global R&D. University of Pennsylvania Dep. Architecture, 701 Graduate Masters level Design Studio. Instructor: Robert Stuart-Smith, TA: Musab Badahdah. Cemex Global R&D: Vice President: Davide Zampini, Workshop Leaders: Alexandre Guerini, Carlos Enrique Terrado, Valentina Rizzo, Matthew Meyers. Students: Yunzhuo Hao. (Celia), Jonghyun Park, Qiaoxi Liu (Josie), Daniel Hurley, Khondaker Rahman (Onie), Mariana Righi, Hasan Uretmen, Ruochen Wang, Yuwei Wang, Yuhao Wu, Yi Dazhong, Xing Zhang, Shiling Zhong.*

LavaSim (2019): *Autonomous Manufacturing Lab, University of Pennsylvania. Research Director: Robert Stuart-Smith. Research Assistant: Patrick Danahy*

Image Credits

Figures 3.1-3.11. *materialAgencies: Washington University, Masters Level Fabrication Studio shared student project. Instructors: Robert Stuart-Smith & Robert Booth.*

Figures 3.12-3.15, 3.17a, 3.18, 3.19. *Robert Stuart-Smith Design Ltd.*

Figure 3.16. *Robert Stuart-Smith Design Ltd. Photograph by Robert Stuart-Smith.*

Figure 3.17b. *Robert Stuart-Smith Design Ltd. Photographs from Shutterstock.*

Figures 3.20-3.24. *University of Pennsylvania Weitzman School of Design, 701 Studio Stuart-Smith & Cemex Global R&D. Photos by Musab Badahdah and Cemex Global R&D.*

Figures 3.25. *University of Pennsylvania Weitzman School of Design, 701 Studio Stuart-Smith & Cemex Global R&D.*

Figures 3.26–3.27. *University of Pennsylvania Weitzman School of Design, 701 Studio Stuart-Smith & Cemex Global R&D. Photos by Cemex Global R&D and Musab Badahdah, University of Pennsylvania.*

Figures 3.29–3.30. *Autonomous Manufacturing Lab, University of Pennsylvania Weitzman School of Design. Director: Robert Stuart-Smith.*

Figures 3.31–3.33. *University of Pennsylvania Weitzman School of Design, 701 Studio Stuart-Smith & Cemex Global R&D. Photos by Cemex Global R&D and Robert Stuart-Smith, University of Pennsylvania.*

References

Ahlquist, Sean, Dillon Erb, and Achim Menges. 2015. "Evolutionary Structural and Spatial Adaptation of Topologically Differentiated Tensile Systems in Architectural Design." In *Artificial Intelligence for Engineering Design, Analysis and Manufacturing: AIEDAM*, 29:393–415. Cambridge University Press. doi:10.1017/S0890060415000402.

Bennett, J. 2010. *Vibrant Matter: A Political Ecology of Things. A John Hope Franklin Center Book.* Duke University Press.

Braidotti, R. 2013. *The Posthuman.* Wiley.

Bryant, L R. 2011. *The Democracy of Objects. New Metaphysics.* Open Humanities Press.

Cemex. 2020a. "Concreto Poroso: Pervia." https://www.cemexmexico.com/productos/concreto/marcas-de-concretos/pervia.

———. 2020b. "Resilia Reinforced Concrete." 2020. https://www.cemex.co.uk/resilia-hyper-performance-reinforced-concrete.aspx.

Cuito, A, and C Montes. 2003. *Gaudi: Complete Works.* Cologne: DuMont.

Dourtme, Stella, Claudia Ernst, Manuel Jimenez Garcia, and Roberto Garcia. 2012. "DigitAl PlAster: A PrOTOTyPICAL DESIgN SySTEM." In *ACADIA 2012 - Synthetic Digital Ecologies: Proceedings of the 32nd Annual Conference of the Association for Computer Aided Design in Architecture*, 2012-October: 217–30. ACADIA.

Fisac, Miguel, and De La Sota, Alejandro. 2013. *Miguel Fisac & Alejandro de La Sota: Parallel Views.* Edited by Alberto Anaut. Madrid: La Fabrica/Museo ICO.

Glancey, J, and F Léglise. 2013. "Kokkugia: Architecture Numérique." *L'Architecture D'Aujourd'hui* 397 (Sept-Oct): 66–71.

Hooke, R. 1676. *A Description of Helioscopes and Some Other Instruments. Cutlerian Lectures.* T.R.

Hubeli, Roger, and Julie Larsen. 2016. "THINNESS: Collaboration Design and Industry." In *2016 TxA Emerging Design + Technology Conference Proceedings: 3-4 November 2016, Held During the Texas Society of Architects 77th Annual Convention and Design Expo in San Antonio, Texas*, edited by Kory Bieg, 8–22. San Antonio: Texas Society of Architects.

Kudless, Andrew. 2011. "Bodies in Formation: The Material Evolution of Flexible Formworks." In *Integration Through Computation - Proceedings of the 31st Annual Conference of the Association for Computer Aided Design in Architecture, ACADIA 2011*, 98–105.

Latour, B. 2009. *Politics of Nature.* Harvard University Press.

Libet, Benjamin. 1999. "Benjamin Libet Do We Have Free Will?" *Journal of Consciousness Studies* 6 (8).

Lloret, Ena, Amir R. Shahab, Mettler Linus, Robert J. Flatt, Fabio Gramazio, Matthias Kohler, and Silke Langenberg. 2015. "Complex Concrete Structures: Merging Existing Casting Techniques with Digital Fabrication." *CAD Computer Aided Design* 60. doi:10.1016/j.cad.2014.02.011.

Malafouris, L, and C Renfrew. 2013. *How Things Shape the Mind. Cognitive Science/Archaelogy.* MIT Press.

Malafouris, Lambros. 2008. "At the Potter's Wheel: An Argument for Material Agency." In *Material Agency.* doi:10.1007/978-0-387-74711-8_2.

Menges, A. 2012. *Material Computation: Higher Integration in Morphogenetic Design.* Edited by A Menges. Architectural Design March/April 2012, Profile No 216. Wiley.

Piker, Daniel. 2013. "Kangaroo: Form Finding with Computational Physics." *Architectural Design.* doi:10.1002/ad.1569.

Schanz, S. 1995. *Frei Otto, Bodo Rasch: Finding Form: Towards an Architecture of the Minimal.* Axel Menges.

Stuart-Smith, Robert. 2018. "Approaching Natural Complexity - The Algorithmic Embodiment of Production." In *Meeting Nature Halfway*, edited by Marjan Colletti and Peter Massin, 260–69. Innsbruck University Press.

Tibbits, S. 2016. *Self-Assembly Lab: Experiments in Programming Matter.* Taylor & Francis.

West, M. 2016. *The Fabric Formwork Book: Methods for Building New Architectural and Structural Forms in Concrete.* Taylor & Francis.

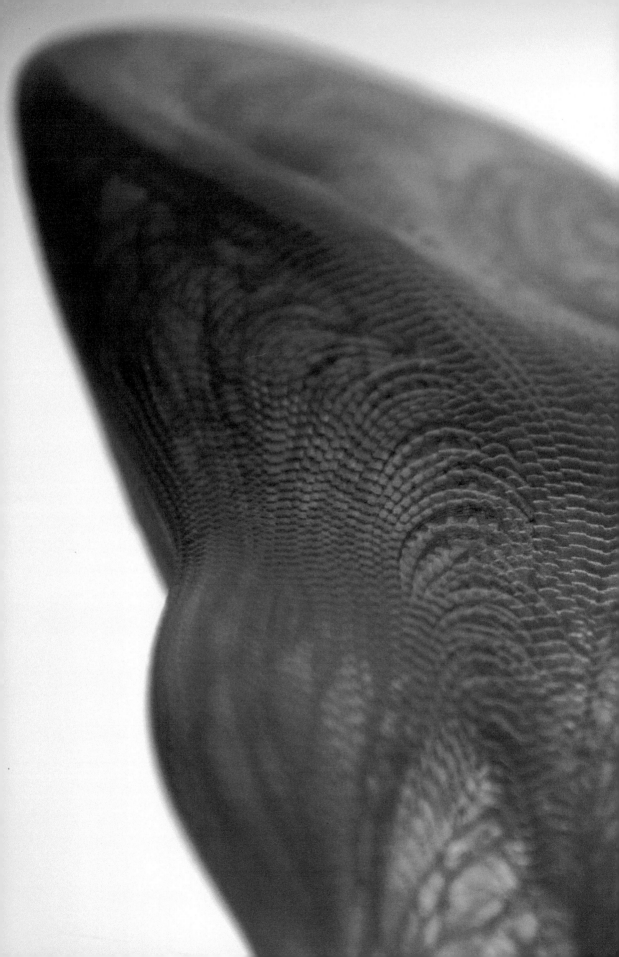

◀ ◀

**4.1 National Art Museum
of China (NAMOC)**
*Photograph of a polyjet 3D
print scaled architectural
model now in the permanent
collection of Gallerie Frac,
Orléans, France.*

4 | Autonomous Agents
Encoding Design Intent Within Matter

While architecture has been traditionally designed through explicit acts of drawing or modelling, there are practical and creative potentials in engaging with generative design methods due to their inherent ability to synthesise several design considerations. In the previous chapter, material dynamics, manufacturing processes, and acts of assembly were employed to creatively impact design outcomes. As generative processes, these approaches were relatively narrow in scale and scope yet managed to align design opportunities to production workflows. For larger architectural projects, additional design criteria must be considered. These include immaterial considerations not easily resolved through material processes, that are better addressed using computational methods. This chapter reframes architecture as a complex, polyvalent organisation of matter, and explores the use of multi-agent computation in 3D graphical space to realise both spatial and aesthetic design intent. Agency is assigned to several types of 3D computer graphics media which consequently generates a design character that is inherent to the specific media and designed behavioural processes.

Architecture and Computation

Architecture, as a field, has developed a diverse range of computational approaches to design. Some notable contributions include research from Nicholas Negroponte's Architecture Machine Group (Negroponte 1972) and John Frazer's *Evolutionary Architecture* (Frazer 1995) in the 1960s, or the playful reprogrammable architecture of Cedric Price with Gordon Pask, John and Julia Frazer in the 1970s (Price et al. 2016). A digital architecture movement that commenced in the 1990s[1] also significantly impacted the profession. Historian and critic Mario Carpo has classified this period into two distinct transformations he calls 'turns' (Carpo 2017). Within Carpo's first turn, several protagonists were interested in ontological aspects of architectural design and explored the generation of form. Peter Eisenman's *The Formal Basis of Modern Architecture* (Eisenman 2018), Greg Lynn's *Animate Form* (Lynn, Arts, and Kelly 1999), and the algorithmic work of Cecil Balmond (Balmond, Smith, and Brensing 2007) offer three different perspectives from this period[2] which have influenced this chapter's work.

Carpo's second digital turn commenced early this century and is characterised by high-resolution designs that make use of big data, additive manufacturing, or robotic fabrication. According to Carpo, the second turn moves away from architecture's traditional reliance on perspective projection and instead focuses on voxels, point clouds, and discrete element assemblages, among other techniques (Carpo 2017, 99). Early projects in this chapter were developed during a period contextualised by a shift towards algorithmic means of design[3] that arguably facilitated the emergence of Carpo's second turn[4]. While some of the projects operate through point clouds (particles), many also leverage curve and surface 3D graphics media that are not often associated with the second turn. This media was engaged due to its ability to render more variable character than voxels or discrete assemblies typically achieve, particularly in physical outcomes that cannot be infinitely scaled and subdivided as readily as their digital counterparts can (e.g., Figures 4.11, 4.12, 4.15. 4.22). In the last few chapters of this book, the use of computer vision or embodied and situated robotic approaches to design diverges a little from the second turn. Projective methods are reintroduced, while projects also engage with physical materials and environments through behavioural design approaches. These later chapters potentially signify an additional (digital or post-digital?) turn invested in material engagement[5].

In architectural design terms, Carpo's first 'turn' prioritises formal continuity and architectural wholes over individual parts, while the second's focus on discrete computation and assemblages creates field-like or aggregated compositional effects without defining wholes. In both cases, architecture creates what Antoine Picon has described as a *"crisis of scale and tectonic"* (Picon 2010, 124–33). The work discussed in this and the next chapter exhibits continuity at times but also embodies high-resolution field-like conditions[6]. Regardless of whether Picon's crisis is averted, many of the projects, except for the first, appear different when viewed from various distances (e.g., Figures 4.4, 4.11, 4.12) and cannot be reduced to a uniform formal condition. Instead, a more complex architecture is sought that engages in a multiplicity of design considerations and effects.

▼

4.2 Little Collins St Baths
a) Rendered perspective section through an atrium/circulation space, and b) an exterior perspective. A new multi-manifold surface mediates between circulatory and programmatic spaces within an existing building shell.

a

b

▶

4.3 Little Collins St Baths
Zcorp plaster 3D print scaled architectural model of the multi-manifold inner surface. Manifolds on the top of the model operate as skylights into the atrium space shown in 4.2.a. a) Overall perspective, b) Close-up of the surface's aesthetic character.

a

b

Architecture is Polyvalent Matter

Historically, architectural design has operated through a diverse set of principles that have encompassed spatial-functional planning, building form, geometry, tectonic assemblage, manufacturing, as well as social, material, and phenomenological effects. Although designers may prioritise different aspects, the effectiveness of their design intent is best evaluated in each materialised outcome. Even architectural drawings reveal these different approaches to organising space and matter. Le Corbusier's Plan Libre, for instance, demonstrates a distinct strategy compared to Loos' Raumplan, which operates primarily in section (Risselada et al. 2008). Michelangelo's Laurentian Library in Florence employs a Mannerist approach to matter at the scale of walls, columns, and reliefs, creating a playful contrast to the adjoined Brunelleschi's Basilica San Lorenzo. While their floor plans may differ, it is in the elemental scales of composition and form that their major distinctions arise (Wölfflin et al. 1966). These examples offer just a sample of the diverse range of design agendas and scales of operation that architecture embodies. Ultimately, an architect's success can be attributed to how effectively they integrate their design intent into architectural matter.

In *Notes on the Synthesis of Form*, Christopher Alexander discusses the importance of managing interrelated constraints in design projects. He argues that a design's success depends on how well it reflects a diagram of its multiple criteria, which he calls a *"constructive"* diagram (Alexander 1964, 61, 88). While the need for a design solution to formally reveal its constraints is debatable, it certainly embodies multiple design concerns. A design outcome that integrates several intended performances and effects could be considered polyvalent. Examples of a polyvalent architectural matter can be drawn from Robert Venturi's seminal book *Complexity and Contradiction in Architecture* (Venturi and Scully 1977), where he discusses the role of hybrid objects that act as 'doubly-functioning' elements. For Venturi, 'doubly-functioning' elements can operate in small architectural details such as baroque cornice ornaments that double as window frames,

Autonomous Agents

architraves that transition into arches or in larger spatial and structural elements (Venturi and Scully 1977, 39). Venturi Scott-Brown's Venturi House project features one such object, combining a stair, chimney, fireplace, and front entry into a single hybrid (Venturi and Scully 1977, 312). Although these examples demonstrate polyvalence, they adhere to architectural syntax, employing combinatorial and transformational principles. Such approaches can produce variety but not generate new material alternatives. In combining features, the chimney-stair remains recognisable, rather than materialising into something altogether different. Despite the design's success, a more integrated approach might have forgone part of the object for something stranger, yet more effective at achieving multiple goals. What if the chimney-stair had to negotiate 20 design criteria – how could it then maintain its legibility? To do so would require the creation of a more complex composite architecture that negotiated all design constraints with a greater level of abstraction in order to achieve a more tailored, specific outcome.

Towards a Complex Architecture

As architecture embodies numerous complementary and competing design criteria, a design approach with more sensitivity to the diverse requirements and effects demanded of the built environment would be beneficial. A means to produce greater levels of polyvalency within architectural matter would enable a more 'complex' architecture. In *Complexity and Contradiction in Architecture*, Venturi argues for a complexity derived from contradiction, juxtaposition, anomaly, and iterative modification (Venturi and Scully 1977). In the context of evolutionary biology, however, complexity is altogether quite different. Scientist Richard Dawkins describes something as complex if it performs a useful function proficiently without a simple explanation for how its heterogeneous and interrelated parts formed, and where there is little chance that a random assemblage would perform equally well (Dawkins 1986, 6–9). Meteorologist Edward Lorenz, known for 'Chaos Theory', provides two other descriptions of complexity; the first ascribes complexity to patterns produced through sensitive dependence (where subtle changes in their initial conditions radically alter results), and the second, to things that visually appear to require a lengthy series of instructions to create, yet can be produced through simple mathematical functions (Lorenz and Hilborn 1995, 24, 167). Both Dawkins and Lorenz are

▼
4.4 National Art Museum of China (NAMOC)
Close-up photograph of a polyjet 3D print scale model revealing an organisational relationship between structure, skin and the building volume.

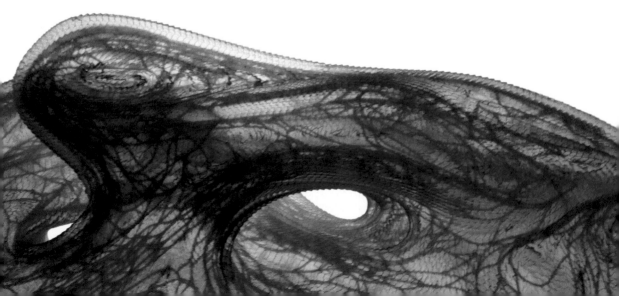

describing a scientific notion of complexity, that arises from non-linear processes of formation. Non-linear complexity does not solely modify or combine established orders; it is capable of creating singularities – unpredictable new orders – and is thus inherently a creative endeavour.

Computer algorithms offer one means to explore such complexity; however, most are rather abstract. Although a vast diversity of creative computational processes have been developed, few offer a suitable means to address the spatially complex nature of architectural design. Many are either; (a) non-spatial, (b) algorithmically fixed to a particular function that is unable to support substantial modification whilst remaining effective[7], (c) morphodynamic[8] and unable to support the strategic networking and weighting of internal morphogenetic[9] design criteria, or (d) indexical – unable to break from an a priori spatial matrix of relations. Standing outside of these limitations, agent-based modelling offers a morphogenetic computational approach to design that is well suited to spatial problem-solving and holds great promise in supporting greater levels of polyvalence in architectural designs.

Agent-Based Models
Agent-based computer models are used in scientific and engineering fields to explore solutions to complex problems ranging from economic modelling (Holland and Miller 1991), to traffic congestion or pedestrian flow analysis and optimisation (Ball 2005). The film and computer game industries also use agent-based models to develop character behaviour or complex crowd simulations. Operating through a time-based simulation, these algorithms program individual or collectives of "agents" – autonomous entities capable of making decisions based on their individual interactions and perceptions (Šalamon 2011, 112). Agent-based simulations model complex scenarios through the programming of simple rules of interaction – essentially event-based rules. In scientific experiments, rules are developed as hypotheses to understand the mechanisms behind complex phenomena, while in entertainment applications, agent models provide efficient means to demonstrate diverse and intelligent improvised individual or collective actions and effects. Despite operating through relatively simple rules, agent-based models are difficult to predict, and difficult to program without feedback. Agent rules are only executed in response to local events, whose likelihood of occurrence is not only impacted by external triggers, but also by an agent's earlier decisions, or in the case of multi-agent systems, interactions with other agents. The parallel execution of multiple rules per agent can make it even more difficult to discern the relation between any individual rule and long-term system-scale effects unless there is some form of regular feedback that provides a programmer with some perception of system-wide behaviour.

While an agent-based stock trader is an a-spatial abstract construct, informatics doctorate Tomas Salamon asserts that spatial problems are well suited to agent-based methods due to their provision of visual feedback that registers behaviour over time, providing visual cognition of the workings of the system (Šalamon 2011, 101). Agent-based models are therefore more easily engaged for architectural design applications that involve problem formulation and resolution in spatial terms compared to alternative non-spatial uses. They are particularly well-suited to generative algorithmic design, robot manufacturing, and collective robotic construction applications presented in this book.

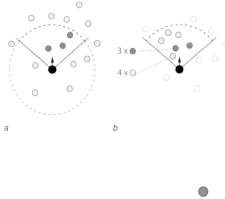

a   b

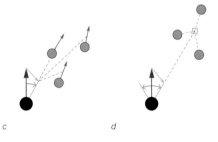

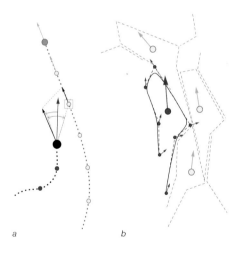

c   d

a   b

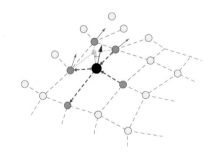

c

Computer scientist Craig Reynolds developed an approach to programming movement in autonomous spatial characters that partly inspired much of this work. Reynolds' agent is an abstract 'vehicle', a particle that moves with a velocity over a series of iterations within a simulation while making autonomous decisions to steer in different directions (Reynolds 1999). One variation of Reynolds' agents, called 'Boids', interact with one another through three agent-to-agent rules: cohesion (an attractive force), separation (a repulsive force) and alignment (orientation). Each of these rules calculates a vector that is proportionally weighted and summed to provide a single steering vector that may be added to an agent's acceleration to inform its next motion step. With these rules, Reynolds' Boids demonstrate swarm behaviour, similar to that seen in schools of fish or flocks of birds, despite no rules explicitly describing this collective behaviour (Reynolds 1987a). Flocking is an 'emergent' phenomenon that arises through individual rule-based interactions that cause the agents to self-organise.

Self-Organisation and Emergence

The term self-organisation is used in several scientific fields to distinguish order that arises through internal interactions in a system (Camazine et al. 2003, 7). Beyond bird flocking or fish schooling, slime moulds, termites, and chemicals self-organise (Johnson 2012). Where pattern formation is produced, results are often considered to be "emergent". Emergence describes a discernable order not explicitly defined by the rules of a system. It is an unpredictable, qualitative effect characteristically common in self-organisational systems (Holland 2000). This unpredictability was understood as early as 1889 through Henri Poincaré's study of the gravitational forces of orbiting planets. Poincaré's 3-Body problem not only impacted the study of celestial mechanics and dynamics theory, it also described one of the core properties of a self-organised system; that a system comprised of three or more elements in a networked series of relationships can produce unpredictable behaviour (Barrow-Green 1997, 240). While unpredictability may be seen as a problem in many scientific fields, it is potentially valuable to designers. The use of bottom-up, event-based rules enables a systemically open means to spatial problem-solving that can be adapted to variable circumstances, providing a flexible approach to generative architectural design.

4.5 Behaviours:
Agent-to-Agent
a) radius: agent only
perceives other agents
within a radius range
and field of view vision
cone, b) agents change
state in response to
agent state quantities in
range, c) steer to align
with other agents, d)
steer to seek or avoid
other agents positions.

An initial foray into self-organisation was explored in the experimental design project *Little Collins St Baths* (2002). A generative process was developed to distribute a thermal bath's program within an existing building shell. The design methodology resulted in a porous topological surface that formally organised the intimately packed program around a continuous labyrinth of circulation and atria. While such an elaborate method was perhaps unnecessary for planning logistics in such a small building, this twenty-two year old project also demonstrated the generation of a unique design character that was integral to the overall design methodology and was not easily achievable by other means at the time[10] (Figures 4.2, 4.3). A particle-spring physics simulation was set up within the building shell and specified a series of conflicting criteria for the location and dimensions of each program (each represented by a particle) along with spring-based physical connections that drew certain programs together and others apart. The simulation was then tasked with resolving program locations and adjacencies into a self-organised, near-equilibrium condition[11]. Although the particles were not autonomous agents that could support more differentiated local interactions, their differentiated constraints, mass, volume, and connections provided bottom-up influences that informed the overall simulation outcome. Exploring more specific architectural designs, however, requires design intent to be encoded into the interactions of such particles, providing them with the ability to make autonomous decisions or to completely change their behaviour in specific circumstances.

Multi-Agent Design

4.6 Behaviours:
Various Multi-Agent
a) steer to seek, avoid
or align to trails other
agents leave in their
wake, b) agent body
seeks, avoids and aligns
to other agent bodies.
Agent body control point
agents seek to snap to
adjacent body control
points, c) surface agent
node seeks, avoids,
aligns to other agent
nodes, and is influenced
by spring forces
connecting it to adjacent
agent nodes. Agents and
springs can be deleted or
added based on nearby
agent state populations,
or an agent's number of
spring connections.

More complex task-oriented behaviours can be observed in the self-organisation of ants or termites. These social insects collectively navigate, congregate, and build emergent structures through local interactions. Scientists like Guy Theraulaz and Eric Bonabeau have also demonstrated the algorithmic nature of such self-organisational systems within multi-agent simulations (Camazine et al. 2003). Instead of simply imitating natural systems, the development of similar self-organisational models as design methodologies enables designers to define specific agencies that are relevant to architectural design goals. By granting individual agents' autonomy in carrying out assigned design tasks, collective behaviour can achieve specific goals that address multiple design criteria. Resulting emergent, negotiated design outcomes must then be evaluated on how well they embody design intent, and where necessary, software modified and iteratively re-run until satisfying results are achieved.

The adaptive and bottom-up nature of agent-based model enables the creative negotiation of multiple design criteria alongside the consideration of practical constraints. Even more interesting, however, are the integral aesthetic implications of such self-organised systems. Agent-based models can generate complex emergent order whose character and heterogeneity are associated with their embodiment of bottom-up, local negotiations. Such properties are difficult to achieve within explicit or parametric models governed by global, top-down rule sets (Roland Snooks and Stuart-Smith 2011). Motivated by both pragmatic and aesthetic possibilities, a multi-agent approach to design[12] was developed by Kokkugia[13] to operate in 3D graphical space and further developed in Robert Stuart-Smith Design[14] and the Autonomous Manufacturing Lab[15] to engage in additional digital, material and robotic informed behaviour (Stuart-Smith 2011, 2016; Snooks 2021). Developed across several projects and applications over two decades, these multi-agent systems are encoded with architectural design intent. Agents operate in a virtual

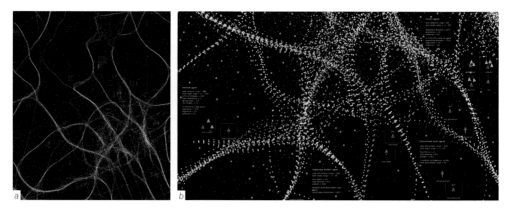

▲
4.7 Swarm Urbanism
Melbourne Docklands
Urban Master
"Algorithm" (in lieu
of master-plan). a)
Docklands plan, b)
Close-up plan.

3D environment or are embodied to operate on an individual robot or swarm of robots. A subset of this work is presented in this chapter that instructs collectives of thousands of virtual agents to self-organise in order to generate architectural designs in 3D graphical space (Figures 4.5, 4.6). Each of these projects involved the development of software where specific autonomous event-based decision-making capabilities were coded to operate on narrowly focused design problems that informed part of an architectural design. Beyond the encoding of architectural design criteria, this work is distinguished relative to others' contributions to multi-agent design by the continuous development of a conceptual and methodological design approach that leverages agency across diverse media to produce project-distinctive formal and organisational character[16].

Developing Distinctive Design Character in 3D Graphics Media

Today's 3D modelling software is built around a small number of computer-graphics object types such as the point, line, curve, polygon mesh, subdivision or nurbs surfaces that, together with industry-established software modelling operations, arguably impact a project's design character, in the same way that watercolour, acrylic, or oil paints affect the visual and material character of a painting. Artists, however, frequently pursue concepts and methods that operate more specifically through that media, demonstrating artistic expression within the scale, distribution and gesture of painting, as can be seen in movements such as Pointillism, Fauvism, Abstract Expressionism, and others. In parallel to addressing architectural design criteria within projects, this design research also explored the development of bespoke design character, through the embedding of design agency within 3D graphical media. 3D graphical space operates as an additional domain in which design is realised through behavioural expression.

All forms of 3D graphics media are defined mathematically and embody distinguishable formal character. A design developed using 3D graphics must operate at a sufficiently high resolution (or agent population) to ensure it is not only organisationally resolved, but also embodies aesthetic character beyond that described by the mathematical description of the media alone. To illustrate the point, a voxel array or nurbs curve with millions of data entry points will exhibit more unique character derived from the data, than voxels or curves based on five data points, where stepping or curvature provided by the underlying media is more apparent.

Resolution or population provides design specificity. Developed methods thus operate at sufficiently high resolutions to register organisational and aesthetic impact beyond 3D graphic media's in-built character.

In this body of work, the production of a complex architecture was sought that integrates multiple performances and effects, intrinsic to a behavioural process of formation. In each project, multi-agent software was coded to suit project-specific design intent and employed within design approaches that involved various combinations of iterative computer programming, explicit 3D modelling, and a wide variety of additional design operations. While several projects defined performance-orientated problems within their methods such as structural or geometric criteria (discussed in the next chapter), the projects presented in this chapter were selected for their ability to produce specific design character within 3D graphical space.

Stigmergic Motion (Autonomous Points)

Particle systems, comprised of a cloud of point objects, are often utilised in time-based computer simulations to approximate the material behaviour of fluids, cloths and solids (soft and rigid body simulations). Typically morphodynamic, such particles have an internally specified mass, while their motion is governed by shared external environmental influences such as gravity or obstacles. If these particles are assigned agency they operate as a morphogenetic system, able to individually determine a unique course of action and, together, produce collective behaviour. Steering behaviours similar to those developed by Craig Reynolds (mentioned earlier)(Reynolds 1987b) provide an internal set of rules that enable each particle agent to produce a distinct motion trajectory (Figures 4.5, 4.6), and can give rise to collective formations that are complex in order, extremely temporal and highly sensitive to perturbation (Figure 4.19). In order to compute more meaningful design formations through agent motion, spatiotemporal considerations must be extended beyond the fleeting moment each agent makes a decision. A principle discovered in the communication and coordination of social insects such as termites and ants, *stigmergy* enables such an extension to space-time (Grassé 1959). Termites leave pheromones in saliva within building material they deposit that becomes information upon which subsequent termite building is adapted to. Similarly, each ant leaves information in the environment in pheromones along the path it traversed that indirectly influences the route of other ants without requiring direct coordination between the ants (Heylighen 2016). In both cases, this spatiotemporal extension in information exchange enables these bottom-up systems to self-organise into efficient outcomes, into towering termite structures

▼
4.8 Swarm Urbanism
Example of stigmergic route finding. Agents are seeded on the right side at a single location at frequent intervals and wander until they find a blue dot whilst seeking and aligning to previously left agent trails. Over time agents create a connective route to the dot by following each others trails.

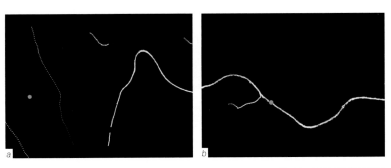

a   b

Autonomous Agents

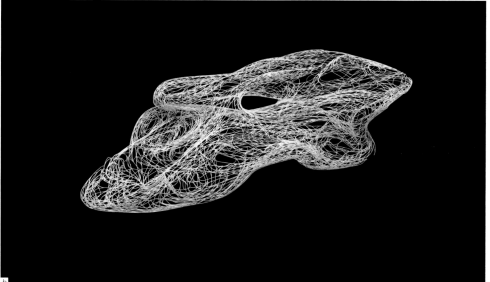

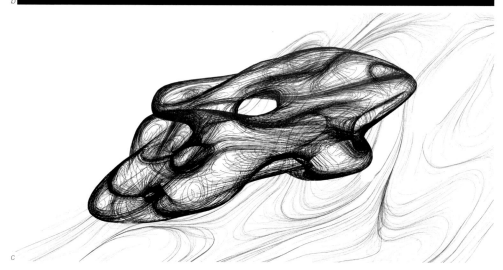

4.9 NAMOC
Integrated design systems: a) glazed envelope, b) glazing structure, c) agent trajectories for building massing and landscape generated through different agent behaviours.

or minimum-detour ant circulation routes. Stigmergy is leveraged in this chapter's multi-agent design methods to enable organisational outcomes to be computed across a specific design space that can persist as either a dynamic stability or near-equilibrium condition discussed towards the end of this chapter in the section "Non-linear Formation".

A project for the Melbourne Docklands, *Swarm Urbanism*, provides an illustration of how simple variations of agent can be used to compute different spatial conditions. The project coincided with an urban redevelopment project under construction in Melbourne that extended the central business district into a disused port territory. Operating more as a provocation than a fully fledged proposal, *Swarm Urbanism* involved the programming of a series of urban systems developed through the local interactions of autonomous agents whose activities gave rise to the self-organisation of urban infrastructural and circulatory networks. This abstract time-adaptive urban development model employs both agent space-packing and path-finding methods (Figure 4.7).

4.10 NAMOC
Three design variations in the glazing structure's network connectivity derived primarily from variations in agent perception parameters (radius and field of view) that influence agent interactions.

Particle-agents with cohesive and avoidance behaviours spatially pack the urban environment to create and maintain a site-specific density that is adaptive to other urban changes over time. A second agent-based system formed a distributed infrastructural network, where the interconnection of agent nodes is achieved using a series of springs whose forces influence each agent's position. Agents had the ability to make or break connections to regulate local network connectivity. A circulation strategy was also developed as a minimum-detour network. Similar to Frei Otto's urban circulatory networks, a minimum-detour network minimises the total length of all routes while connecting necessary points of interest (Schanz 1995). In *Swarm Urbanism*, however, a minimum-detour network is generated through an agent trail-following approach (Figure 4.8) that is conceptually similar to how ants create their highly trafficked circulatory routes by a *Stigmergic* process mentioned just before (Grassé 1959). Applying this approach to agent-based programming, agents leave a trail of data points in their wake that other agents seek and align to. Over simulated time a diverse range of wobbly trails that connect points of interest consolidate down to a reduced, more direct series of routes to compute a minimum-detour circulatory network. Together these different swarm systems interact with one another to operate as a multi-agent ecology that self-organises to arrive at an urban condition, shifting urban design from notions of the master plan to that of the master algorithm. While the former defines a concept that must be rigidly followed over decades, the latter is an adaptive system that is perhaps more aligned with the reality that

4.11 NAMOC
Photograph of a 3D print scaled model. The glazing distribution demonstrates a chaotic order with turbulence that is not a subdivision of the building's geometry yet is integral to its formal and topological design. All glazing panels are of identical size and shape, with varying degrees of overlap and alignment to a glazing structure visible on the interior.

a

b

c

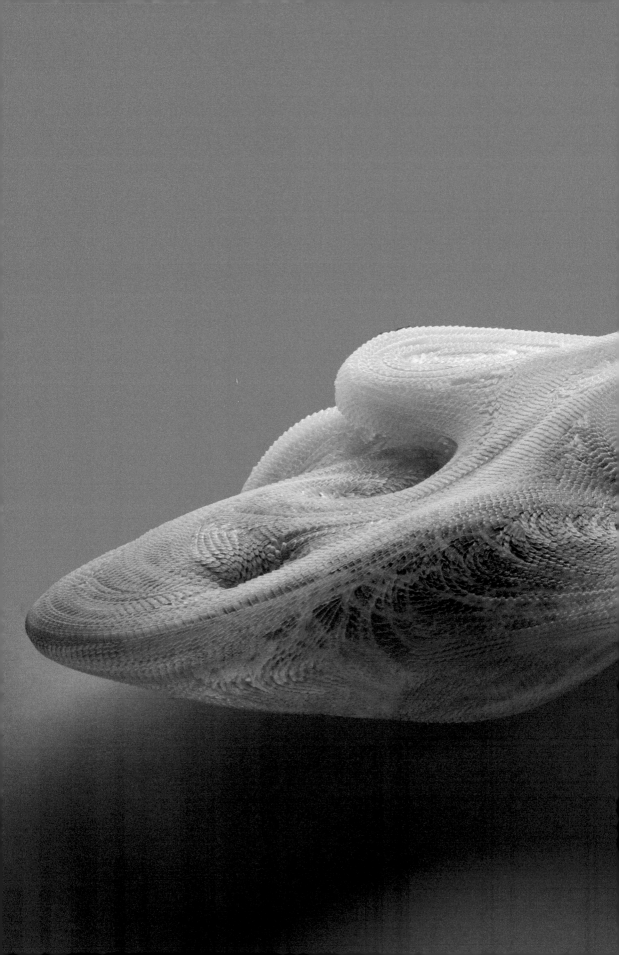

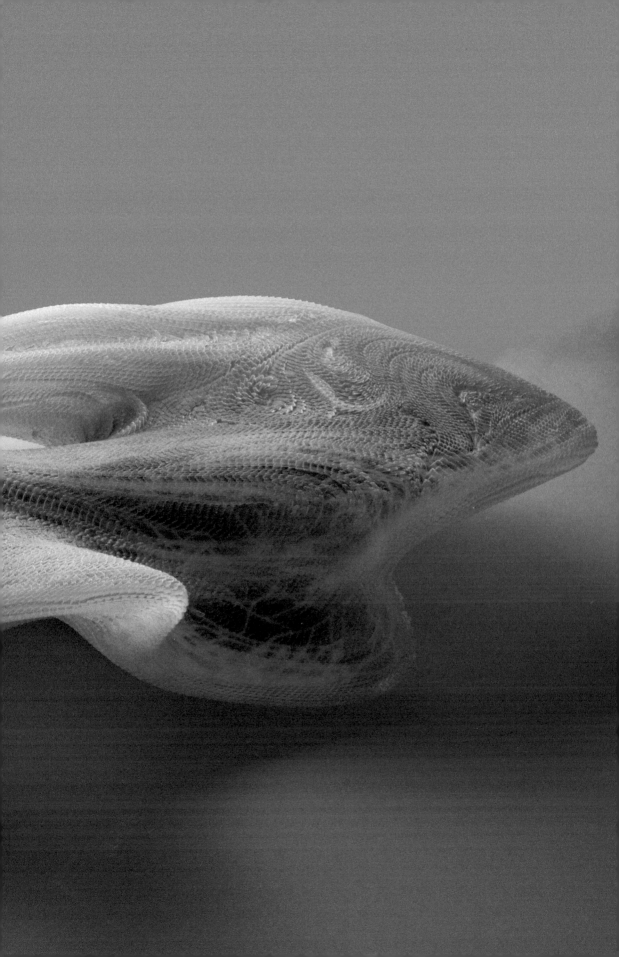

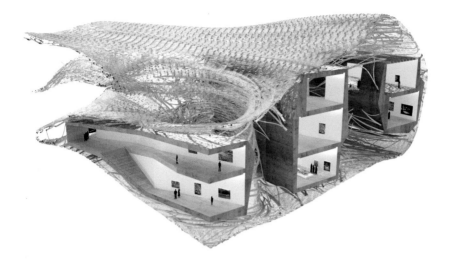

◀

4.12 NAMOC

Landscape podium tiling represented from three different scales of observation. The multi-agent-generated order offers different levels of expression when seen from different distances.

▲

4.13 NAMOC

3D chunk sectional perspective. The glazed exterior envelops a series of interior concrete gallery spaces.

cities are dynamic entities, constantly in flux. Although *Swarm Urbanism* is relatively abstract, similar agent-based motion principles were developed for specific building design methodologies.

A proposal for the *National Art Museum of China (NAMOC)* (Brayer and Migayrou 2013; Glancey and Léglise 2013) involved a collaboration with Studio Zhu Pei, for the design of a building 'cloud' that visually floated above a landscape, and intended to resonate with Chinese culture's numerous affiliations with clouds. Seeking a building and landscape that would appear to embody the turbulent dynamics of clouds, it was hoped that a chaotic order would also reconcile the large scale of the site with the scale of individual building elements, allowing an iconic 240-metres-long building to engage in smaller-scale orders and effects (Figure 4.9). To achieve this, a multi-agent system was tasked with developing a glazed envelope and glazing structure that wrapped a series of long concrete gallery volumes, in addition to defining the geometry and tiling of an expansive hard landscaped exterior space (Figures 4.9, 4.10). Each of these design aspects was developed through the programming of different variations of agent-based motion that exhibited turbulence. For each of these design outputs, a cloud of particles autonomously moved across a 3D model during a time-based simulation employing similar stigmergic behaviours to those described in the *Swarm Urbanism* project to produce a series of motion-trajectory curves (Figure 4.9c). Each agent's motion was influenced by the proximity and alignment of neighbouring agents and their trails, perceived within a range and field of view (Figure 4.5), with agents constrained to move primarily along the surface of a 3D massing. This 3D massing was explicitly and iteratively modelled in parallel to iterative programming and testing of the multi-agent simulation. Both the agent simulation and explicit model commenced independently from one another with their activities converging over several iterations to produce a design character that is the product of their combined influence.

While both the building and landscape exhibit turbulence, behavioural character in each was differentiated through variations in their multi-agent rules (Figures 4.9c, 4.12). These algorithmic changes were aesthetically

◄
4.14 NAMOC
Close-up of glazing
panels and glazing
structure. Photograph
of 3D print scaled
architectural model.

motivated, whilst also accommodating design considerations related to intended material and economic constraints (e.g., it is easier to employ variable geometry in landscape paving than glass panels relative to their material, assembly and labour costs). The landscape trajectories converge more readily, defining a series of landscaping ridges, while the building exterior registers this same turbulence through a field of overlapping glass panels that are distributed along agent trajectories that exhibit more alignment and less convergence. This organisational difference is also reflected in the geometric translation of both tiling systems; the landscape agent-behaviour required less alignment, producing more irregular hard-landscaping tiling where every tile is geometrically unique – based on a Voronoi pattern distributed along the agent trajectories control points (Figure 4.12). In contrast to this, the glazing system consists of panels of identical size and shape, distributed and orientated along agent trajectories, creating a heterogeneous effect solely by variations in the distribution, overlap, and orientation of panels (Figures 4.1, 4.14).

NAMOC's glazing panel distribution is rather unique due to the fact it was not derived from a geometrical subdivision of the building geometry (Figure 4.11). Most buildings maintain an indexical relationship between façade tiling and building geometry, except where tiling might be naively juxtaposed on building geometry without much integration effort (Figure 3.18). Divergent from both approaches, in *NAMOC* there is an intrinsic relationship between the panel organisation and the building geometry. However, this relationship is not indexical; instead, it embodies more local adaption and heterogeneity, arrived at through agent-based autonomous programming (Figure 4.4). Directly beneath this glass envelope, a glazing structure was generated by a third multi-agent system tasked with moving along an interior offset of the building envelope, to generate a series of curves from agent motion trajectories that connect to the glazing system and each other (Figure 4.11). The topology of this curve network was controlled through the regulation of each agent's range and field of vision, which impacts their ability to perceive and move closer to or connect with an adjacent trajectory curve (Figure 4.10). *NAMOC's* design was developed through an iterative back-and-forth between multi-agent programming alongside 3D modelling activities. Explicit modelling together with generative algorithmic design enabled the project to explore an aesthetic relationship between form and different means of behavioural organisation in 3D space, each driven by architectural design concerns. The project demonstrates a complex integration between multiple building systems through related processes of behavioural formation. The tectonic concept, however, was only arrived at through the designer's directives in producing co-related sets of agent simulations that did not necessarily embody co-related constraints. Agency can also be directly encoded within a tectonic concept to better interrelate different material or spatial constraints.

Agent-Bodies (Autonomous Curves)
To explore a tectonic agency, experiments were also undertaken into processes of self-organisation that could produce field-based assemblages, similar to how a collective of ants can form bridge-like structures through the interlocking of their bodies. Curve object "bodies" were tasked with moving and connecting with each other to form larger, geometrically varied field organisations. These agent-bodies operate with a behavioural hierarchy, first as a body in motion that seeks, avoids, and aligns to other bodies, and secondly, with limbs that also grasp hold of adjacent bodies which also modifies the geometrical configuration of each body in the process. Early

▶

4.15 BodySwarm
Two different design
outcomes (a,b). A
complex emergent order
arises from the non-
linear spatial interaction
of identical agent-body
curves that seek, avoid,
align and snap to each
other's profiles.

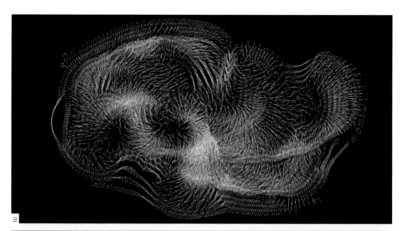

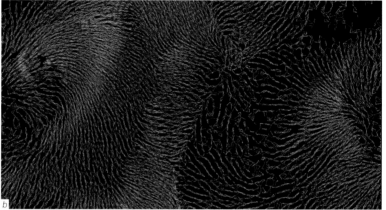

examples of agent-bodies research include Kokkugia's *Swarm Matter* and *Developmental Fields* projects. *Swarm Matter* unified a series of bodies as an isosurface mesh and revealed local formations of symmetry amongst a larger differentiated field, while Developmental Fields explored how a branching agent-body could create a tectonic system, operating as a bundled network structure and roof hybrid field-condition (Figures 4.16, 4.17).

In Developmental Fields, agent-bodies could be instanced with a variable number of branching limbs and connect with adjacent bodies, sharing at minimum, one polyline segment. Each polyline represented a centre-line for a diamond-profiled extruded element, with the shared polyline segment enabling these extrusions to be bundled into a larger variable space-frame-like structure. Although the agent-based algorithm was relatively simple, a far more complex algorithm had to be coded to calculate which side an extruded profile should be located off-centre in order to not intersect with adjacent profiles throughout its different bends to connect with adjacent agent-body profiles. The bundled 3D structure was conceived to be partially align to embed in a roof envelop surface, while also projecting into space either side of the surface at times to create more structural and visual depth.

A similar, more intricate field-based order was developed in a series of video and print artworks acquired for the FRAC Centre's permanent collection called *bodySwarm* (Brayer and Migayrou 2013). In *bodySwarm*

(Figures 4.15, 4.18), simple curve geometries swarm in 2D graphical space. Each curve shifts in position and orientation throughout simulated time using steering behaviours like those described earlier (Figures 4.5, 4.6). Each curve's control points were also tasked to snap onto adjacent curves' control points. Although each agent's identical curve-based geometrical bodies were simple with few control points, the collective organisation, overlapping, and snapping of curves to adjacent curves produced substantial spatial and topological variation. This is due to the non-indexical networked locations of the individual curve agents, which were free to adjust their position and orientation over time in relation to each other, self-organising at the scale of both bodies and limbs. Together, these explorations into agent-body field formations demonstrate the potential for a vastly heterogeneous tectonic order. It was envisaged that with the possibilities for variable manufacturing using robotic fabrication, processes like rod or tube bending would enable Developmental Fields' geometrically differentiated parts to be relatively economically produced.

Body Topologic (Autonomous Surface)
While *NAMOC* and bodySwarm are spatial, they operate as field organisations and do not engage in the formation of space, but rather its organisation or formal articulation. Space is essentially infinite until bounded by closed-manifold geometry (a piece of paper or a cup are open-manifold geometries as they have exposed edges while a sphere is a closed-manifold geometry). Once bounded, beyond a surface's morphological properties (e.g., form, scale, proportion, curvature), its topological genus (number of holes) might impact its spatial organisation or structural efficiency (see Chapter Six). Designs by architectural firms such as UN Studio have arranged program, circulation, voids and other spatial considerations, around topological concepts (Van Berkel

4.16 Developmental Fields
a-b) agent-body bundled structure, c-d) agent-body bundled structure is embedded into a roof surface.

4.17 Developmental Fields
a-b) a 3D print scale model chunk of an agent-body bundled space-frame and roof envelope.

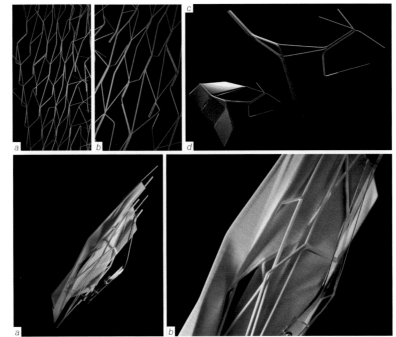

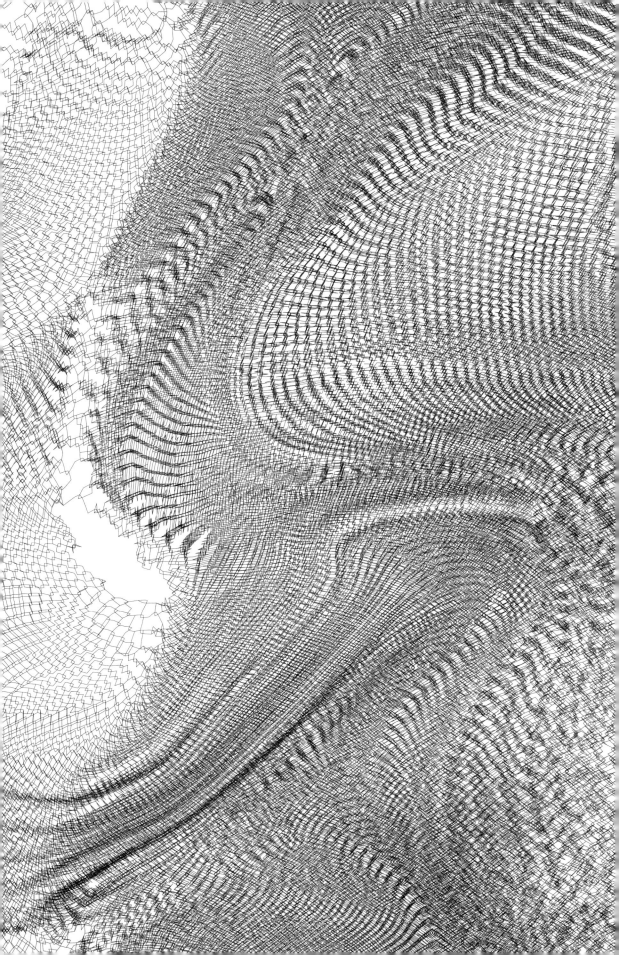

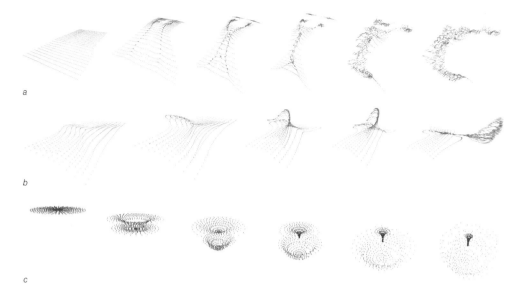

a

b

c

◀◀

4.18 bodySwarm
A single time-frame from a bodySwarm simulation. The simulation commenced with identical curve agent-bodies distributed in parallel rows before beginning to self-organise into a more complex pattern that exhibits an almost wrinkled fabric-like character.

▲

4.19 bodyTopologic
(a-c): Initial multi-agent topological formation studies. Agent formations are volatile and too easily perturbed.

and Bos 1999). To pursue this generatively, however, the organisation of space within architectural design requires engagement with the formation of multi-genus topology.

To explore the generative design of topology, a multi-agent surface formation software was developed called *bodyTopologic* (Stuart-Smith 2011). Inspired by the fascinating topological transformations depicted in Stuart Pivar's *On the origin of form: Evolution by self-organisation* (Pivar 2009), in *bodyTopologic*, complex behavioural relations trigger topological events. Instead of using a mathematical description of topological events like Rene Thom's catastrophes (Thom 1994), *bodyTopologic* takes inspiration from developmental biologist Luis Wolpert's perspective. Wolpert suggests that complex organisations are better understood through a set of generative instructions rather than descriptive results (Wolpert et al. 2007, 26, 28). In *bodyTopologic*, surface formation is controlled by intensive relations, allowing a design to emerge from a series of adaptively programmed events.

Initial experiments were developed that sought to define topological events through collectives of freely moving agents that were loosely affiliated with each other. While these experiments demonstrated immense potential, they also highlighted the fragility of self-organised topological orders if such orders were only dynamically stable and did not result in a near-equilibrium condition that might persist and resist perturbation. Each study produced an emergent outcome momentarily and then dissolved into lower levels of dispersed or clustered order. Further, as these experiments assigned agency to points, there was no easy way to transition a simulated outcome to a modifiable surface topology post-simulation that could be carried forward in design proposals. To overcome these challenges, it was determined that agency should be assigned to a surface itself, allowing the surface to self-organise and modify its topology while ensuring a simple manifold surface outcome could be achieved. In *bodyTopologic* a population of agents self-organises to create an emergent multi-genus topological surface outcome.

A *bodyTopologic* simulation commences with a user-defined input, as simple as a point in space or a surface, and then proceeds to modify the input's topology through bridging, stitching, tearing and hole-creation, all initiated by local event-based rules. Although structured in a similar way to a polygon mesh geometry that comprises mesh faces, vertexes and edges, a *bodyTopologic* surface is not indexically organised, and was not coded with any mesh library dependencies. Each object type is defined in a Javascript™ class[17], with each object-oriented instance having knowledge of other object instances it is attached to (object connections can be made and broken at any point during a simulation). Decision-making takes place primarily within vertex-class agent instances that operate similarly to agents in previously discussed projects. Each vertex is capable of autonomous decision-making that governs its motion and connectivity to other agent vertexes, springs and faces, while face edges are programmed as springs (with elasticity defined by rest length, stiffness and damping values)[18] (Figures 4.5, 4.6).

Topologic surfaces transform over simulated time through events perceived locally by each vertex and spring. Vertex-agents are aware of whether they are located on the edge or inside of a surface simply by querying their local connections (a vertex on a naked edge will not have more than two or three connected springs or vertexes). Based on such local perceptions, vertexes change state, with each state programmed to behave in different ways that include: seeking, avoiding or aligning with other vertexes, make or break spring connections of varying elasticity, or by modifying surface topology through either stitching (add spring

▼

4.20 BodyTopologic
(a-d): A series of formation studies arising from the orchestration of event-based rules that causes changes of state in a multi-agent topological body.

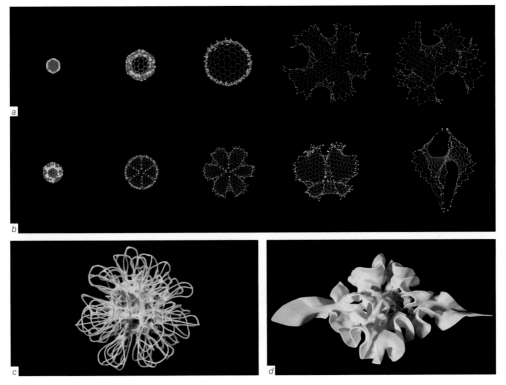

4.21 Busan Opera
The design's figure cantilevers over the water with tendon-like features that extend back into the massing. This formal articulation was not 3D modelled but emerged as an integral part of a topological process of formation.

connections + faces), tearing (remove spring connections and faces), or bridging (by deleting two opposing faces and creating a network of springs and faces to fill in the void). A *bodyTopologic* surface is therefore to some degree self-aware, but without global knowledge, it operates as a collectively self-organising system. *bodyTopologic* received its first road-test in a short workshop in the Design Research Laboratory Masters program at the Architectural Association School of Architecture in London[19]. Students were provided with the algorithm and tasked with defining a very simple initial vertex and spring configuration, and local event-based rules that would inform topological transformations over time. The goal of the short workshop was to develop animated surfaces with characteristics integral to their process of formation. While results were quite simple (Figure 4.20), the degree of formative potential the algorithm demonstrated proved a success and provided confidence that *bodyTopologic* was ready to be used in a design project.

A proposal for the *Busan Opera House* Competition was developed using *bodyTopologic* (Figure 4.21). A simple design template (a rectangular prism with an initial vertex and face arrangement) was defined together with a few anchor positions to ensure that the generated design would remain anchored to the site (Figure 4.22). Whilst design outcomes were unknowable prior to running the simulation, iterative changes allowed for incremental design improvements. The project involved hundreds of simulation runs, each with various amendments to either the initial design template or algorithmic behaviours. Together, these placed the self-organisational system into a scenario in which it had to solve its relations to arrive at a design proposal. With designer intuition operating throughout these activities, both pragmatic and qualitative design criteria were honed throughout the process.

A complex topological design emerged from this surface agency, where the local interaction of agent-based vertexes caused a range of surface conditions to arise and trigger further change. The design result was then output as a 3D polygon mesh. While part of its design character is determined by this media, its topology, form and creases play a significant role. Not manually modelled, they emerged from its non-linear process of formation. Event-based rules triggered local changes in surface formation, resulting in a unique, muscular, tendon-like surface character (Figures 4.23, 4.24). In an explicitly modelled surface, the silhouette might be modelled first, with smaller design features added as an act of design refinement. In contrast, in the Busan project, smaller features were produced by local agent interactions and were consequential to global surface formation. Encoded in the *bodyTopologic* software's multi-agent surface's autonomous behaviours, this local surface character was intrinsic to the design's overall process of formation.

Non-linear Formation

In both *NAMOC* and *bodySwarm*, a time-frame is extracted from each simulation as a design outcome. These frozen moments capture a computed order that emerges after the perpetual simulation settles into a dynamically stable state. The concept of *"dynamic stability"* was developed by Lynn Margilis and James Lovelock to describe a persistent complex order exhibited by the earth's ecosystem, despite it remaining in constant change (Lovelock 2010), precariously 'on the edge of chaos' (Lovelock 2007). Differing from this, the *Busan Opera House* project's design simulation undergoes radical changes until it reaches a final topological condition where only subtle formal oscillations continue. This is characteristic of a spring system where the springs approach their rest length and exert minimal force, coming to near rest. In the *BodyTopologic* software, both agent and spring motion arrive at a near-equilibrium state. In these projects, when a design solution was considered adequately computed if it had reached either a dynamically stable or near-equilibrium state. As both conditions arise from a non-linear process, they are unpredictable prior to simulation despite involving no random properties. Given this unpredictability, some may question the extent of a designer's authorship when leveraging such autonomous processes or their ability to critically evaluate and develop design proposals using such methods.

Design Agency

There are scenarios where computational approaches to design might be considered to rob the architect of some agency. Examples include; when a designer is unable to view or modify compiled software, or where designers force their formal design outcomes to fit a specific

▼
4.22 Busan Opera
A bodyTopologic generative design simulation gains spatial, formal and topological definition over time. The initial conditions for a simulation (a) operate as an abstract "design template" that together with bodyTopologic event-based rules, determine a design outcome (b).

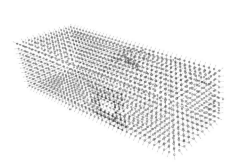

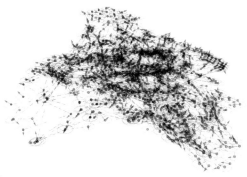

a

b

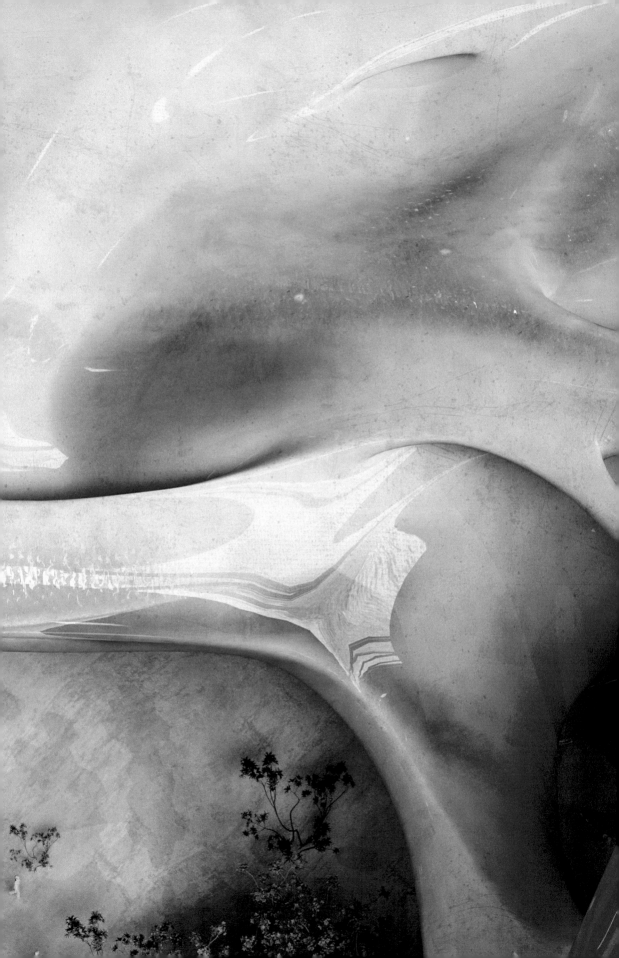

92

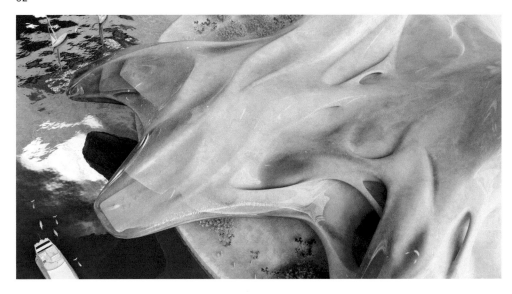

◀◀

4.23 Busan Opera
*Close-up view from
above. The unique
muscular and tendon-
like surface character
emerged from a process
of topological formation.*

algorithmic output (such as the use of a branching algorithm to design a branching building). The former is similar in limitations to explicit 3D modelling within established software programs, although the designer might potentially have even less choice or operational awareness. The latter is simply a non-creative engagement with software programming, where an algorithm is treated like an object trouvé. The projects discussed in this chapter offer a counter-narrative to these scenarios, where designers develop an intuition for the behaviour of authored autonomous systems, enabling uninhibited creativity in the same way as in any other design medium.

▲

4.24 Busan Opera
*The muscular character
of the design arises
from surface tension
within a body Topologic-
generated formative
process. Conceptually
envisaged to be clad in
ceramic and glass, the
cantilevered extensions
to the massing transition
to a translucent state
over the water.*

Working with agent-based programming, design operates through the orchestration of events. The designer constructs scenarios and contingencies, and designs via semi-autonomous behaviours. Outcomes are controlled through bottom-up programming and occasional top-down interventions (modelled or software-coded), which inform subsequent alterations to programmed behaviours. This iterative development operates in a similar manner to the potter's wheel mentioned in the previous chapter. The execution of an agent-based simulation provides feedback to the designer, whose subsequent decisions adjusting the value of variables or geometrical inputs are a reaction to indirect outcomes of their earlier decisions. Design is not the outcome of one simulation, it is the result of a compounded, complex set of interactions where the designer in some form collaborates with, and adapts to, their programmed autonomous entities over numerous iterations. This does not preclude the designer from intervening with explicit 3D modelling or structural code adjustments. In this process, architectural design intent is abstracted to behavioural rules that individually might not appear to bear any relation to architecture, but through their spatiotemporal enactment produce an intended architectural purpose that might operate at the scale of a building's formal massing, its structure, cladding, ornament, or any other aspect of architecture. A designer determines the scale and scope of an autonomous system's activities within each project, and the goals of such programmed behaviour.

Design authorship in this chapter's projects expanded into new territories within 3D graphical space by developing custom agent-based approaches to point, curve, and surface self-organisation. Different types of 3D graphics media were creatively engaged through the orchestration of local event-triggered actions, that enabled design to operate through varying degrees of autonomy. These approaches successfully resulted in a unique organisational and aesthetic character for each project. In the next chapter, this behaviour-based approach is further extended to incorporate structural criteria within emergent design outcomes.

Notes

1. Books by Antoine Picon (Picon 2010), Mario Carpo (Carpo 2011), Branko Kolarevic (Kolarevic 2004), Ali Rahim (Rahim 2000, 2002) and Neil Leach (Leach et al. 2004) and others summarise the 1990s digital architecture movement well.

2. The three examples are not all necessarily discussed by Carpo, but are considered by the author to be aligned with Carpo's first turn in different ways.

3. During the time the author's company Kokkugia was in practice several other practices were also exploring various degrees of computation in design including; The Very Many, Matsys, Aranda Lasch, SJET, Minimaforms, Tom Wicombe Design, Biothing, Supermanouvre, Mesne, and several others.

4. Several of the example projects Carpo includes in his book The Second Digital Turn (Carpo 2017) were produced by former students and employees of either the author or colleagues mentioned above.

5. See Chapter Three for a description of material engagement.

6. "Field conditions" is a term that describes spatial distributions of energy or matter that exhibit order. The term as used in this text is derived from a seminal text by Stan Allen that described a field as being primarily about the space between things and explored a field-based architecture as opposed to an object-based one (Allen 1997).

7. E.g., a branching algorithm can be used to design a branching building or structure but not a sphere.

8. A morphodynamic algorithm uses external constraints rather than internal relations. E.g., a particle system might be shaped by a wind force and a barrier, rather than rules that define inter-particle relations.

9. Unlike morphodynamic algorithms that are determined by external influences, morphogenetic algorithms generate outcomes from rules internal to a system. Genetic algorithms, multi-agent systems, or cellular automata are examples of morphogenetic algorithms.

10. To develop the project, a particle-spring physics simulation was employed together with a Windows DOS-based tomographical software package designed to create detailed isosurfaces from medical scan images (comparable methods were not readily available in CAD software at that time).

11. For more information, see the "Non-linear Formation" section in this chapter.

12. The development of custom multi-agent software programs for architectural design and research in Kokkugia was undertaken in collaboration with Roland Snooks.

13. The experimental design practice Kokkugia Ltd was co-founded by Robert Stuart-Smith, Roland Snooks and Jonathan Podborsek in 2004 in Melbourne and subsequently operated from London and New York until 2012.

14. Robert Stuart-Smith Design Ltd, 2013 – Present. Co-Founding Directors Robert Stuart-Smith and Stella Dourtme (Stuart-Smith 2013).

15. The Autonomous Manufacturing Lab has two research labs, 1) University of Pennsylvania, Weitzman School of Design (AML-Penn). Director: Robert Stuart-Smith, 2) University College London, Department of Computer Science (AML-UCL). Directors: Robert Stuart-Smith, Vijay Pawar (Stuart-Smith 2017; Stuart-Smith and Pawar 2016).

16. Kokkugia pioneered several multi-agent design approaches in the field of architecture and championed a conceptual approach to multi-agent design in research and teaching activities at institutions including: the AA School of Architecture (AA.DRL), Columbia University, University of Pennsylvania, RMIT University, and others. Kokkugia's approach explored the encoding of architectural design intent for specific aspects of design and aimed to produce distinctive character as an emergent outcome of each self-organisational process. Kokkugia also targeted these activities to specific aspects of projects, also integrating explicit modelling at times in an iterative, parallel set of interactions between algorithmic and explicit processes. Several colleagues also developed and deployed multi-agent algorithms for different purposes, including Paul Coates and Claudia Schmid (Coates and Schmid 1999), Theodore Spyropolous (Spyropoulos 2016), Alisa Andrasek (Brayer and Andrasek 2009), Supermanoeuvre (Maxwell and Pigram 2023) and others.

17. Java is an object-orientated programming language that has Class and Class instance object types.

18. Springs were coded based on Hooke's Law. Springs are discussed in more detail towards the end of Chapter 6.

19. BodyTopologic Design Workshop, Architecture Association School of Architecture, Design Research Laboratory (AA.DRL) Studio Stuart-Smith. Teaching Assistant: Knut Brunier. Students: Marzieh Birjandian, Nicholette Chan, Nassim Es-Haghi, Georgios Kontalonis, Leonid Krykhtin, Rana Zureikat, Carlos Luna, Bita Mohamadi, Gilles Retsin, Ashwin Shah, Aaron Silver, Sophia Tang.

20. See 16.

Project Credits

Little Collins St Baths (2002): Robert Stuart-Smith

Swarm Urbanism (2009): Kokkugia. Roland Snooks & Robert Stuart-Smith

National Art Museum of China (NAMOC) (2010): Kokkugia. Design Directors: Roland Snooks & Robert Stuart-Smith. Design Team: Nicholette Chan, J Fleet Hower, Xiaotian Huang, Leonid Krykhtin, Edwin Liu, Josef Musil, Ekaterina Obedkova, Casey Rehm, Gilles Retsin, Sophia Tang. Collaborator: Studio Pei-Zhu Partner: Pei Zhu

Developmental Fields (2010): Kokkugia. Design Director: Robert Stuart-Smith. Design Team: Katya Obedkova, Alicia Namad, Nick Williams.

BodySwarm (2010-3): Kokkugia. Design Director: Robert Stuart-Smith. Design Team: Katya Obedkova, Fellippe Escudero.

BodyTopologic (2010): Kokkugia: Robert Stuart-Smith.

Busan Opera House (2010): Kokkugia. Design Director: Robert Stuart-Smith. Design Team: Isaie Bloch, Nicholette Chan, Katya Obedkova, Igor Pantic, Gilles Retsin, Paola Salcedo, Ashwin Shah, Sophia Tang.

Image Credits

Figures 4.2-3, 4.5-6: Robert Stuart-Smith

Figures 4.3, 4.7-18, 4.20-23: Kokkugia Ltd, Robert Stuart-Smith and Roland Snooks. Photographs 4.1, 4.4, 4.11, 4.14 by Dimitrije Miletic.

Figure 4.20: AA.DRL Studio Stuart-Smith design workshop, Architectural Association School of Architecture. Participant-developed geometry[20] generated using Robert Stuart-Smith's bodyTopologic Algorithm.

References

Alexander, C. 1964. *Notes on the Synthesis of Form. Harvard Paperback.* Harvard University Press.

Allen, Stan. 1997. "From Object to Field." In *AD: Architecture After Geometry, edited by Maggie Toy, 24–31. AD Profiles.* Wiley.

Ball, P. 2005. *Critical Mass: How One Thing Leads to Another: Being an Enquiry Into the Interplay of Chance and Necessity in the Way that Human Culture, Customs, Institutions, Cooperation and Conflict Arise.* Arrow.

Balmond, Cecil, J Smith, and C Brensing. 2007. *Informal.* Prestel.

Barrow-Green, J. 1997. *Poincare and the Three Body Problem. History of Mathematics.* American Mathematical Society.

Brayer, M A, and A Andrasek. 2009. *Biothing, Alisa Andrasek. Collection Du FRAC Centre.* HYX.

Brayer, M A, and F Migayrou. 2013. *Naturaliser l'architecture: Collection Du FRAC Centre. Collection Du FRAC Centre.* HYX.

Camazine, Scott, Jean-Louis Deneubourg, Nigel R. Franks, James Sneyd, Guy Theraulaz, and Eric Bonabeau. 2003. *Self-Organization in Biological Systems. Princeton Studies in Complexity.* Princeton University Press.

Carpo, M. 2011. *The Alphabet and the Algorithm. Writing Architecture.* MIT Press.

— — —. 2017. *The Second Digital Turn: Design Beyond Intelligence. Writing Architecture.* MIT Press.

Coates, Pau, and Claudia Schmid. 1999. "Agent Based Modelling." In *Architectural Computing: From Turing to 2000: ECAADe Conference Proceedings,* edited by A Brown, M Knight, and P Berridge. Liverpool: CAADRU.

Dawkins, Richard. 1986. *The Blind Watchmaker: Why the Evidence of Evolution Reveals a Universe Without Design. National Bestseller. Science.* Norton.

Eisenman, P. 2018. *The Formal Basis of Modern Architecture.* Lars Müller Publishers.

Frazer, John. 1995. *An Evolutionary Architecture.* London: Architectural Association.

Glancey, J, and F Léglise. 2013. "Architecture Numérique: Kokkugia." *L'Architecture D'Aujourd'hui* 397 (September): 66–71.

Grassé, Plerre P. 1959. "La Reconstruction Du Nid et Les Coordinations Interindividuelles Chez Bellicositermes Natalensis et Cubitermes Sp. La Théorie de La Stigmergie: Essai d'interprétation Du Comportement Des Termites Constructeurs." *Insectes Sociaux* 6 (1). doi:10.1007/BF02223791.

Heylighen, Francis. 2016. "Stigmergy as a Universal Coordination Mechanism I: Definition and Components." *Cognitive Systems Research* 38. doi:10.1016/j.cogsys.2015.12.002.

Holland, J H. 2000. *Emergence: From Chaos to Order. Popular Science / Oxford University Press.

Holland, John, and John Miller. 1991. "Artificial Adaptive Agents in Economic Theory." *American Economic Review* 81 (2): 365–71. doi:10.2307/2006886.

Johnson, S. 2012. *Emergence: The Connected Lives of Ants, Brains, Cities, and Software.* Scribner.

Kolarevic, Branko. 2004. *Architecture in the Digital Age: Design and Manufacturing.* Taylor & Francis.

Leach, N, D Turnbull, C Williams, and C J H Williams. 2004. *Digital Tectonics.* Wiley.

Lorenz, Edward N, and Robert C. Hilborn. 1995. "The Essence of Chaos." *American Journal of Physics* 63 (9). doi:10.1119/1.17820.

Lovelock, J. 2007. *The Revenge of Gaia: Earth's Climate Crisis & The Fate of Humanity.* Basic Books.

— — —. 2010. *The Vanishing Face of Gaia: A Final Warning.* Basic Books.

Lynn, G, Graham Foundation for Advanced Studies in the Fine Arts, and T Kelly. 1999. *Animate Form.* Princeton Architectural Press.

Maxwell, Iain, and David Pigram. 2023. "Supermanoeuvre." Accessed October 6. https://supermanoeuvre.com/.

Negroponte, N. 1972. *The Architecture Machine: Toward a More Human Environment.* M.I.T. Press.

Picon, A. 2010. *Digital Culture in Architecture: An Introduction for the Design Professions.* Birkhäuser Basel.

Pivar, S. 2009. *On the Origin of Form: Evolution by Self-Organization.* North Atlantic Books.

Price, C, S Hardingham, Architectural Association London, and Canadian Centre for Architecture. 2016. *Cedric Price Works 1952-2003: A Forward-Minded Retrospective. Projects.* Architectural Association, AA.

Rahim, A. 2000. *Contemporary Processes in Architecture. Architectural Design.* Wiley.

— — —. 2002. *Contemporary Techniques in Architecture. AD Profiles.* Wiley.

Reynolds, C W 1999. "Steering Behaviors for Autonomous Characters." *Game Developers Conference,* 763–82. doi:10.1016/S0140-6736(07)61755-3.

Reynolds, Craig W. 1987a. "Flocks, Herds and Schools: A Distributed Behavioral Model." *ACM SIGGRAPH Computer Graphics* 21 (4): 25–34. doi:10.1145/37402.37406.

— — —. 1987b. "Flocks, Herds, and Schools: A Distributed Behavioral Model." In *Proceedings of the 14th Annual Conference on Computer Graphics and Interactive Techniques, SIGGRAPH 1987,* 25–34. Association for Computing Machinery, Inc. doi:10.1145/37401.37406.

Risselada, M, A Loos, L Corbusier, and J van de Beek. 2008. *Raumplan Versus Plan Libre: Adolf Loos [and] Le Corbusier.* Distributed Art Pub Incorporated.

Šalamon, T. 2011. *Design of Agent-Based Models: Developing Computer Simulations for a Better Understanding of Social Processes. Academic Series.* Tomáš Bruckner.

Schanz, S. 1995. *Frei Otto, Bodo Rasch: Finding Form: Towards an Architecture of the Minimal.* Axel Menges.

Snooks, R. 2021. *Behavioral Formation: Volatile Design Processes and the Emergence of a Strange Specificity.* Actar D.

Spyropoulos, Theodore. 2016. "Behavioural Complexity: Constructing Frameworks for Human-Machine Ecologies." *Architectural Design* 86 (2). John Wiley & Sons, Ltd: 36–43. doi:https://doi.org/10.1002/ad.2022.

Snooks, Roland, and Robert Stuart-Smith. 2011. "Nonlinear Formation: Or How to Resist the Parametric Subversion of Computational Design." In *Apomechanes: Nonlinear Computational Design Strategies,* edited by Pavlos Xanthopoulos, Ioulietta Zindrou, and Ezio Blasetti, 1st ed., 192. Asprimera Publications.

Stuart-Smith, Robert. 2011. "Formation and Polyvalence: The Self-Organisation of Architectural Matter." In *Ambience'11 Proceedings,* edited by Annika Hellström, Hanna Landin, and Lars Hallnäs, 20–29. Boras: University of Borås.

— — —. 2013. "Robert Stuart-Smith Design." https://robertstuart-smith.com/.

— — —. 2016. "Behavioural Production: Autonomous Swarm-Constructed Architecture." *Architectural Design* 86 (2). John Wiley & Sons, Ltd: 54–59. doi:10.1002/ad.2024.

— — —. 2017. "AML-PENN." https://www.aml-penn.com/.

Stuart-Smith, Robert, and Vijay Pawar. 2016. "Autonomous Manufacturing Lab: | University College London, Computer Science." https://www.aml-ucl.co.uk/.

Thom, R. 1994. *Structural Stability And Morphogenesis. Advanced Books Classics.* Avalon Publishing.

van Berkel, Ben, and Caroline Bos. 1999. *Move.* Un Studio & Goose Press.

Venturi, R, and V Scully. 1977. *Complexity and Contradiction in Architecture.* Museum of Modern Art.

Wölfflin, H, P Murray, and K Simon. 1966. *Renaissance and Baroque.* Cornell University Press.

Wolpert, L, J Smith, T Jessell, P Lawrence, E Robertson, and E Meyerowitz. 2007. *Principles of Development.* Oxford University Press.

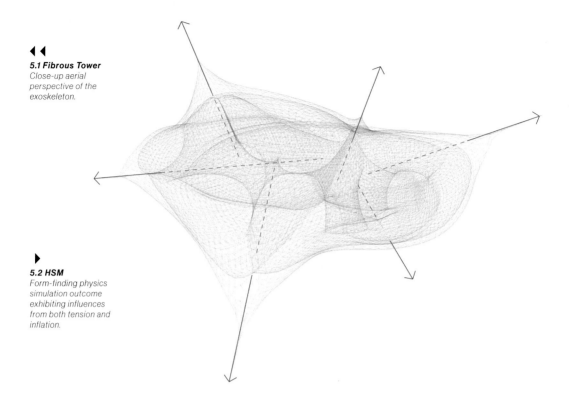

◀ ◀
5.1 Fibrous Tower
*Close-up aerial
perspective of the
exoskeleton.*

5.2 HSM
*Form-finding physics
simulation outcome
exhibiting influences
from both tension and
inflation.*

5.3 HSM
*3D modelled input
geometry, defining a
topological knot with
two short-cut bridges.*

▶
5.4 HSM
*Simulation time-frames
(a-c) of four adjacent
cubes inflated with
different volume and
pressure ratios.*

a   *b*   *c*

5 | Structural Agency:
Encoding Structural Intent within Formative Processes

Formative Processes

In the previous chapter an agent-based approach to design operated within 3D graphical space using point, curve and surface media for the generative design of an architecture that embodied an aesthetic character integral to custom approaches to behavioural formation. It was also argued that agent-based models can be employed to resolve conflicting design criteria through processes of self-organisation by computing either a dynamic stability or near-equilibrium condition. In this chapter, structural criteria are also incorporated within such formative processes.

While diverse relationships between form and structure are evident in vernacular architecture, the design-engineering of structures by Felix Candela (Garlock and Billington 2008), Eladio Dieste (Pedreschi 2000), Eduardo Torroja, Pierre Luigi Nervi (Gargiani, Bologna, and Haydock 2016) and other historically innovative engineers, demonstrates unimaginable thinness and material efficiencies through the employment of mathematically described geometries. More recently Philippe Block, Matthias Rippmann, Masoud Akbarzadeh and others have developed elaborate and efficient shell structures (Rippmann and Block 2013), slabs, and 3D space-frames (Nejur and Akbarzadeh 2021) that are generatively designed using graphics-statics. Although these astonishing works were developed using geometric principles, there is a parallel history to structural design-engineering that arises from non-geometric, physical processes of formation.

Documented approaches to the generative design of structures through physical processes of formation arguably commence with Robert Hooke's development of Funicular structures in the 1600s. By inverting a catenary chain that is suspended under the influence of gravity to operate solely in tension, Hooke was able to design efficient compression-based arches, vaults and domes that mitigated horizontal thrust forces, eliminating the need for large buttresses (Graefe 2020; Hooke 1676). Following Hooke, funicular arches were used to calculate or verify several notable buildings' structures. In the work of Antoni Gaudi though, a more visceral application of the principle operated as a generative design method. Gaudi's design for the Sagrada Familia cathedral was developed through the manipulation of elaborate catenary chain scaled models, where networks of cables were interconnected and differentially weighted using small bags of sand to develop the

a

b

c

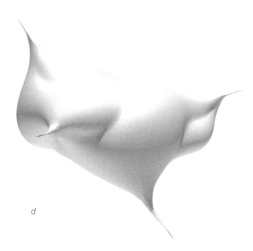

d

cathedral's overall design upside down. After the catenaries came to rest in an equilibrium condition, the form could be inverted to provide an ambitiously tall compression structure without employing the flying buttresses commonly seen in gothic cathedrals (Cuito and Montes 2003; Burry 2013).

Frei Otto also appropriated funicular methods for the design of large-span grid shells, including his wafer-thin design for the Manheim Multihalle (1975) which spans up to eighty metres with a structure made out of 2-inch square (50 x 50mm) sections of wood (Schanz 1995, 140). Almost fifty years on, this lightweight temporary structure remains standing today. During his directorship of the University of Stuttgart's Institute for Lightweight Structures from 1964 to 1991, Otto conducted research that utilised natural processes of formation. He employed material form-finding techniques to identify optimised geometries in small-scale models, which could then be scaled up for use in large-span, lightweight building structures. For example, experiments with surface tension in soap film or lycra scale models resulted in minimal surface geometries that influenced the design of structurally efficient tensile fabric membrane structures, including Otto's renowned design for Munich's 1972 Olympic Stadium (Schanz 1995, 107). From 1958, similar surface tension forces were observed in Otto's "minimum-detour networks". Otto experimented by placing wool threads in different spatial configurations. After briefly submerging the threads in water, they partially adhered together, resulting in a smaller subset of interconnected paths. This method optimised the overall length of a network of curves by consolidating routes that were close to each other. As a result, Otto was able to design distinctive and efficient circulation routes, as well as fascinating branching column structures (Schanz 1995, 68).

Beyond Hooke and Otto, Heinz Isler was developing structural form-finding methods for concrete shell structural designs as early as 1954 (Isler 1980). Similar to Candela or Torroja, Isler's large-span shells were exceptionally thin and structurally efficient. Unlike Candela and Torroja's mathematical approaches, however, Isler developed physical experiments to design his curved shells. Isler's form-finding techniques included the funicular draping of fabric and

◀

5.5 HSM
*Final design outcome:
a) interior membrane, b)
internal walls, c) interlayer
stiffeners between inner
and outer membranes, d)
exterior envelope.*

"bubble shells" – where an inflated pillow of air was constrained by a horizontal frame on its perimeter (Chilton and Isler 2000). Each of these methods involved simulating the forces acting on a substitute material in a scale model to determine the form of a concrete shell design.

Isler and Otto's careers were also preceded and no doubt influenced by, D'Arcy Thompson's 1917 book, *On Growth and Form*, which provided an early study of how physical forces impact the formation of natural phenomena. Thompson demonstrated how surface tension and fluid dynamics can guide the formation of various living and non-living structures. A splash caused by dropping a round pebble into a liquid is compared to a hydroid polyp (Thompson 1992, 70–71), and the formation of a liquid jet with vortexes created by a falling drop of liquid through another liquid is related to different jellyfish (Thompson 1992, 70–75). While Thompson drew similarities between these phenomena, he also acknowledged Galileo's principle of "similitude," which suggests that structures cannot be infinitely scaled up while maintaining their original formal or material proportions (Thompson 1992, 19; Thompson 1915). Thompson observed that the ratio of bone to flesh in land-based vertebrates increases with the animal's size[1], while the skeletal proportions of a porpoise relative to a whale are minimal due to the reduced influence of gravity in water. Thompson also noted that inter-molecular forces have a greater influence than gravity in microscopic examples (Thompson 1992, 20, 37). Similar to the formative processes Thompson describes, Gaudi, Isler, and Otto's work orchestrates physical forces to generate form across different materials and scales. Although their experiments are scaled up dramatically to become buildings, only the form is transferred, with materials and thicknesses adjusted to accommodate the shift in scale.

Gaudi, Isler, and Otto's models were not representational architectural models created to communicate their proposals to clients. Instead, these operative design models utilised a formative process to address relationships between form, space, and structure. Designs were arrived at through negotiations, guided by the physical interactions of mass, gravity, and surface tension within a self-organising material process. Although generative, their approach had limitations. The physical nature of these models made it impractical to incorporate design criteria that could not be physically described. Although such methods can

▶

5.6 HSM
Exterior perspective.

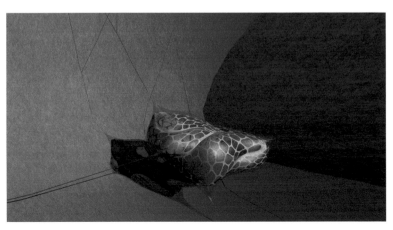

Structural Agency

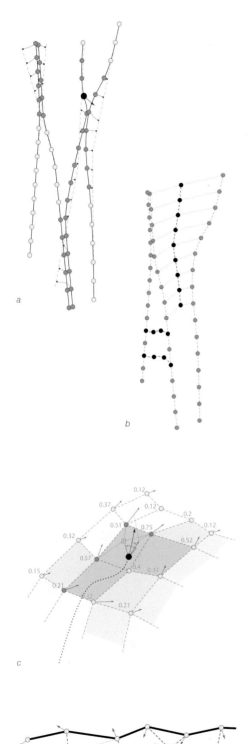

a

b

c

d

now be modelled in computer simulations (Kilian and Ochsendorf 2005; Pigram and McGee 2011), modifying such processes with additional rules or manipulating a result in post-simulation can potentially undermine the physics-generated solution. Thus, even as a computational process, the physical aspect of the approach constrains geometric outcomes. While a designer might iterate and refine input conditions, their inability to radically alter the logic of the process locally without undermining it renders such a computational model more automated than autonomous (similar to the Jacquard Loom or Heron of Alexandria's Automatons mentioned in Chapter One). In contrast, semi-autonomous methods of self-organisation allow for more diverse formative processes to be developed, without limitations on the number of design criteria that can be implemented. Also, with less risk of the method being undermined, such an approach can more easily accommodate intuitive interventions by designers throughout the creative process.

Semi-autonomous algorithmic approaches enable designers to engage in the event-based rules of formation itself (Figure 5.7). Such models have the potential to address not only structure and form but also several other architectural design criteria. These criteria, at times, may take priority over form or structure. The focus of the work discussed in this chapter is not to produce optimal structural solutions, but rather to incorporate structural considerations into design outcomes. Together with a host of other design criteria, structural principles become integral to spatial, formal, environmental, or ornamental design. In this sense, the design process is not characterised by optimisation so much as by negotiation and expression.

Surface Formation and Structure
An early project, *Habitation Sur Mars (HSM)* offers a simple illustration of a process of formation that integrates multiple design criteria. *HSM* was a speculative project for an early Mars research habitation proposal. As with many exoplanet habitat concepts, *HSM* is an inflatable that requires only the membrane (and not the volume) to be transported to the red planet, saving limited transportation payload for other things. Inflatable structures are arguably relatively efficient on Mars where an Earth-like interior pressurisation

5.7 Structural Agency
a) agent curve self-organisation. Curve nodes seek and snap to adjacent curve's nodes, b) additional curves comprised of agent nodes are created between existing curves, either as an infill or bridge condition, c) agent motion is constrained to a 3D mesh structural analysis outcome, steering towards or away from high deflection or utilisation values, while aligning with principle stress vectors, d) section diagram of agent-springs network and its relationship to a roof surface (bold) and building volume on right side. Agents make and break spring connections based on other agents' proximity and spring force magnitudes.

would have considerable resistance against an outside environment with approximately a third of the atmospheric pressure and gravity relative to Earth's. Unlike other proposals, however, *HSM* is suspended above the ground on tensile cables anchored into Mars' rock. *HSM* is thus an inflated and tensile structure (Figure 5.2). Additionally, a topological knot of interior space was proposed that enabled a series of individual rooms to be connected into one continuous circulation path (Figure 5.3). Developing a computational design approach to support this concept with the limited software resources available at the time was challenging (design and animation software were not up to the task twenty years ago). A custom-developed workflow incorporated software from scientific and mathematical disciplines in addition to custom-scripted routines.

HSM was developed through computer simulations that sought to resolve both internal air pressure and surface tension forces acting on a simple 3D-modelled input geometry (Figure 5.4). The input model specified the design's topological configuration, outlining its overall spatial organisation and adjacencies. The multi-criteria simulation process produced a formal outcome that was not structurally optimal, but demonstrated a negotiation of space, form, tension, gravity and pressure. Interestingly, this resulted in an almost muscular sculptural-like surface character within the design, intrinsic to both its inflated and tensile conditions (Figure 5.5). However, similar to Gaudi and Otto's design processes, as *HSM* was developed through a physics-based computer simulation, it did not incorporate autonomous event-based rules that would enable design agency to be locally differentiated. The software *bodyTopologic*, discussed in the previous chapter, addressed this years later. Used for the design of the *Busan Opera House* (see Chapter Four, Figures 4.21-4.24), *bodyTopologic's* multi-agent generative design approach to topological surface formation is more systemically open and can incorporate additional design criteria that can be heterogeneously addressed within a design. Following the Busan project, the *bodyTopologic* method was also extended to include structural analysis feedback and utilised within the design of other projects.

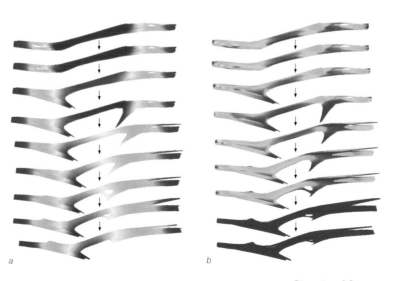

a   b

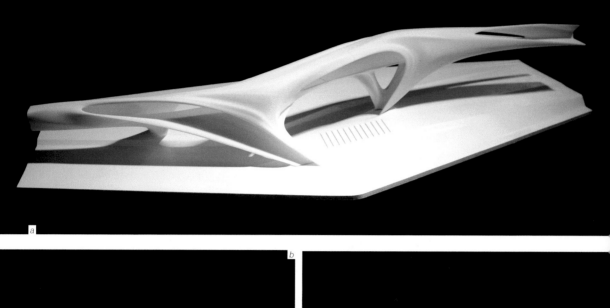

a

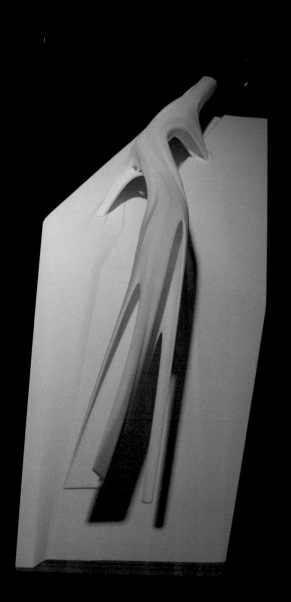

b

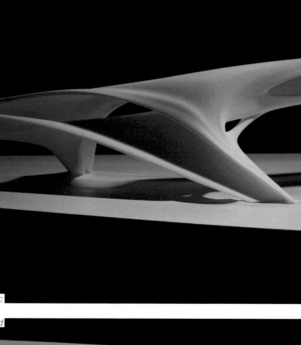

c

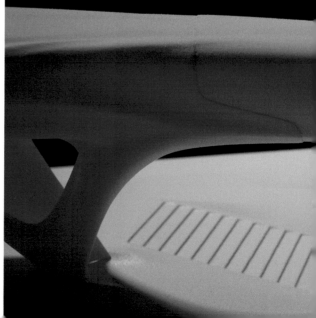

d

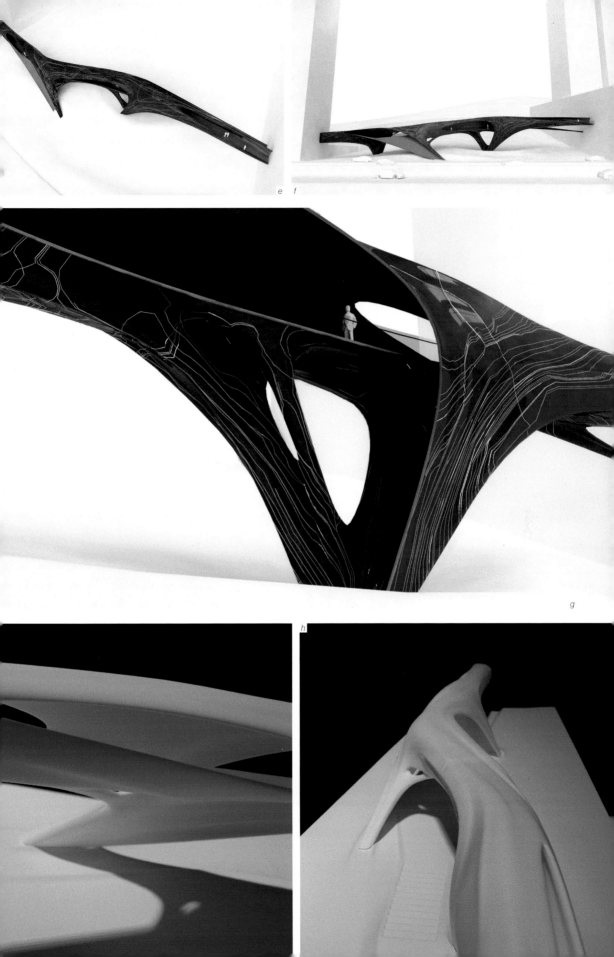

e   f

g

h

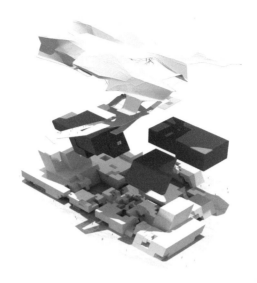

▲▲

5.9 Sao Paulo Bridge
*a-d, h) photographs
of architectural scale
model, e-g) views of
the bridge on site with
connections to adjacent
buildings at different
floor levels.*

◀

**5.10 Taipei
Performing Arts
Center**
*a) Exterior perspective,
b) interior view from top
floor of podium below
roof canopy, c) aerial
perspective. building
shell.*

The *Sao Paulo Bridge* project (Figure 5.9) was designed to be a multi-part, pre-fabricated thin-shell fibre-composite structure[2]. Throughout its design process, as the pedestrian bridge's topological complexity increased, so does its structural efficiency (Figure 5.8). The introduction of additional manifolds and branches creates additional supports while also increasing surface curvature, resulting in a more effective structure. In addition to its topological formation, a series of smaller-scale ridge-like formal transformations further enhance structural performance. Like *HSM*, the *Sao Paulo Bridge* is not an optimal geometry for an ideal bridge. The design integrates a structural solution relative to other design criteria that constrained the project to connect to several specified touch-down points at ground level along with spatial connections to existing buildings across multiple floor levels. The *Sao Paulo Bridge* design approach incorporates structural considerations within its formative process, rendering it intrinsic to its resulting surface character. There are many instances, however, where a building design might not operate primarily through surface, requiring an alternative approach. Another series of projects explores the agent-based formation of networked structures.

Agent-Based Networks

▲

**5.11 Taipei Performing
Arts Center**
*Exploded perspective
showing podium,
auditoria and open
roof canopy above.
Part of the podium is
subdivided and eroded
to create outdoor public
spaces and circulation
routes below the roof.*

Similar to *HSM's* physics-based simulation, the *bodyTopologic* software incorporated a spring system that operates through non-linear dynamics (described in Chapter Four). The use of spring forces[3] in projects designed using *bodyTopologic*, results in a simulation eventually coming to rest in a near-equilibrium surface-based design solution. In *bodyTopologic's* approach, springs are attached to agent nodes, and both springs and agent nodes operate autonomously, making a series of event-based decisions throughout a simulation that produces a surface-based design outcome. A precursor to these projects, the *Taipei Performing Arts Center* proposal (Figure 5.10), utilised a simpler agent-springs approach to develop a 3D space-frame roof system. The roof system operates as an open canopy that hovers above a multi-level

Structural Agency

5.12 Fibrous Tower
Curve-agent self-organisation studies producing a consolidated series of shared paths.

inhabited podium and a series of closed-volume auditoria (Figure 5.11). In the *Taipei* project, the space-frame self-organises through agent spatial decision-making rule sets that also respond to internal force management, where agents' spring connections are made or broken in relation to the magnitude of the attached spring's forces (Figure 5.7d). The roof surface was part algorithmically and part explicitly modelled, and involved several multi-agent simulation runs of the space-frame. While regions of the roof were modelled post-simulation to adapt to a space-frame simulation outcome, it was also possible to fix the position of some space-frame nodes attached to modelled auditoria volumes or to fix previously simulated frame nodes and to re-run the simulation only on parts of the roof for further design refinement. The semi-autonomous aspects of the design approach were thus also iteratively informed by human feedback and explicit modelling (*bodyTopologic* was also developed to support such iterative workflows although most projects utilising it did not use the feature).

5.13 Fibrous Tower
a) Aerial perspective, b) close-up of exoskeleton.

A different network condition is explored in *Fibrous Tower* (Figure 5.13). A concrete exoskeleton supports a column-free flexible office space on each floor. The exoskeleton operates as both structure and ornament while also providing sufficient porosity to enable exterior views and daylighting to the interior (Figure 5.14). In *Fibrous Tower*, two offset layers of a vertically orientated and interconnected network form a spatial exoskeleton. An excess of elements works in concert with one another to create a networked structure with redundancy in individual elements, providing more than one route for forces to pass through the structure. The inner and outer layers of the network are offset from one another, except for in one region where they delaminate to produce additional thickness and provide an interstitial outdoor terrace (Figure 5.15). The project was inspired by Frei Otto's minimum-detour networks mentioned earlier. In contrast to Otto's physical experiments, however, *Fibrous Tower* was algorithmically designed in 3D graphical space.

5.14 Fibrous Tower
Cast-in-place construction strategy: a) formwork assembled from two sides, b) in-situ concrete is poured, c) load-bearing concrete exoskeleton result.

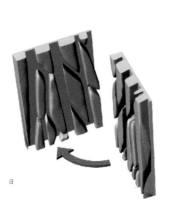

c

b

a

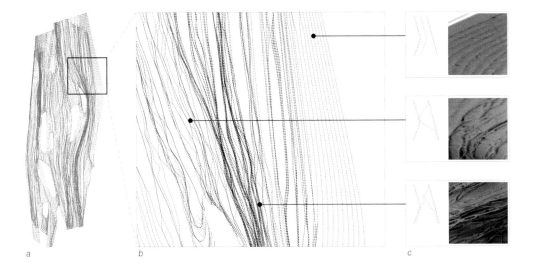

a

b

c

◀

5.15 Fibrous Tower
Perspective section.
Exoskeleton delaminates
to provide interstitial
exterior space.

▲

5.16 AirBaltic
Multi-agent simulation
of roof/ceiling surface:
a) reflected ceiling plan
of roof, b) close-up of
agent-node strands,
c) emergence of three
different structural
strategies in result.

▶▶
5.17 AirBaltic
Landside exterior view
of Departures entrance.

Fibrous Tower's design was developed through autonomous interactions between the control points of a series of curves distributed over the tower exterior (Figure 5.7a). Each curve's control points had agency – to seek out and move towards the closest locations on adjacent curves. Iteratively updating their position relative to one another in simulated time, the curves self-organise into a more interconnected state that also reduces their overall composite length (Figure 5.12). Unlike Otto's physically constrained and homogeneous process, computationally, there are more possibilities to control the scale of networking activity and the number of curve paths a network reduces to. The range of vision[4] of curve control point "agents" can be extended or shrunk to permit only close-quarter or far-ranging self-organisation activities, while the number of curves permitted to consolidate adjacent to one another can also be capped, with agents encouraged to seek out other options if a curve path is oversubscribed by too many other curves. Agent behaviours can also be differentiated for different solar or urban orientations, or in relation to different interior conditions. Collectively, these behaviours enable differentiation in the number, size and orientation of façade openings. Although the translation of the curve network into a volumetric 3D model is also accomplished through a series of scripted routines, design agency operates in the self-organisation of the network of virtual centre-line curves.

The *Taipei Performing Arts Center* and *Fibrous Tower* projects are networked structures, determined through virtual agent-based centrelines that operate with different rule sets to produce different forms of self-organisation. Both commenced from a unique set of initial conditions correlated to their intended application. The *Tapei* design involved an initial random spatial distribution of nodes in the location of a roof that would alter position and degrees of connectivity (making or breaking spring connections between nodes), while certain nodes were fixed from the outset in relation to the project's design model. *Fibrous Tower* commenced with a series of relatively straight, crossing curves arranged in the plane of explicitly modelled interior and exterior envelope surfaces. In both projects, network connectivity varies across the designs, and is locally determined by a semi-autonomous, architectural-scale process of self-organisation. Notably, the translation of simulation outcomes into a volumetric surface geometry

Structural Agency

5.18 AirBaltic
*Interior view. Skylight
and escalator leading up
to departure gates.*

was also arrived at by different techniques, tailored to divergent material and construction considerations. Although both *Tapei* and *Fibrous Tower* addressed structural principles through agent-based network formation that could be refined in dialogue with structural engineers, neither project responded directly to structural analysis data. The following projects extend this approach to incorporate structural analysis feedback in-the-loop and also offer a more topologically free geometrical translation of an agent network that oscillates somewhere between both surface and strand-like conditions.

Surface-Strand Formations

The 2010 *AirBaltic Terminal* competition for Riga Airport in Latvia, involved a collaboration with engineering firm BuroHappold's structural engineering and airport planning groups. The Terminal's architectural brief was addressed through an explicitly designed volumetric surface model that defined spaces, vertical supports and a ceiling scape relative to the airport's vast number of planning constraints. Due to airport programme constraints and architectural design intent, this designed surface was not structurally optimal. For instance, the design benefited from being column-free at ground level although this created large horizontal structural spans. This span could not be realised with an optimal structural form such as a tall arching funicular shell as the airport's second-floor level sat directly above. In response to such strategic considerations, a generative algorithmic design approach was able to improve overall structural efficiency by defining a high-resolution material organisation relative to the predetermined building geometry.

The roof-surface design was abstracted to a series of strands – agent-based linear chains of interconnected nodes (Figure 5.16). While each node-agent was able to move towards other nodes similar to curve edit-point agents in *Fibrous Tower*, they were also able to make or break connections to each other (within or between strands, see Figure 5.7b). Each individual agent iteratively improved its structural fitness by making and breaking connections with neighbouring nodes in response to their own local structural deflection data without consideration of the overall building's structural performance (Stuart-Smith 2011). Structural data was obtained iteratively at each time-step in the simulation. At the commencement of the simulation, the roof strands oscillated rapidly as each agent selfishly sought to improve their structural performance, until over time, the overall roof organisation eventually settled into very minor oscillating movements, reaching a near-equilibrium condition.

The resulting strand organisation is emergent, and curiously, embodied several different structural typologies within its differentiated order, none of which were explicitly defined in the algorithm or initial 3D modelled inputs (Figure 5.16). The strand formation was interpolated as a single isosurface mesh[5]. Strands near one another became unified as a singular volumetric mesh surface, while strands further apart were independent tubular meshes. The embodiment of the strand organisation within the interpolated mesh resulted in variations in smooth and ribbed shell surface in regions, through to truss and space-frame-like conditions elsewhere (Figure 5.17). These diverse orders arose within a process of formation as a direct response to local structural data, rather than through a global analysis or forming process that might operate on the whole design at once. Due to the high degree of articulation in *AirBaltic's* resulting high-resolution roof, the design arguably embodies a structural ornament whose character is intrinsic to its custom generative process.

Situated somewhere between *AirBaltic* and *Sao Paulo Bridge*, *Tendinous Bridge* (Figure 5.19) is part fibrous network, and part surface (Figures 5.20, 5.22). The design methodology is also a hybrid. A surface formation method (similar to *Sao Paulo Bridge*) was first employed to determine the overall arched and branching tubular surface of the bridge. This surface design was structurally analysed. The design's material organisation was then determined by a multi-agent simulation. Agents' behaviours adapt to local structural deflection and principal stress data as they move along the surface, with their motion trajectories[6] creating

▼
5.19 Tendinous Bridge
Aerial perspective of bridge spanning river.

▲

5.20 Tendinous Bridge
*Close-up of design's
surface-strand hybrid
character.*

a network of strands (similar methods are discussed in more detail in Chapter Six). Like *AirBaltic*, these trajectories autonomously build connecting links to one another in relation to structural data (Figure 5.7b). While *Tendinous Bridge's* design methodology operates through a combined generative approach to surface formation and material organisation, the design outcome is embodied within a single isosurface mesh (similar to airBaltic) that interpolates the agent motion trajectory strands and their associated connecting links as one volumetric surface. The initial surface model is only indirectly embodied due to each agent's adherence to the surface model throughout their simulated motion sequence. Where agents' trajectory strands are sufficiently apart, a grain is expressed in the mesh that appears fibrous, while regions where strands are closely packed are interpolated as a single monolithic surface. The project's name is derived from the design's bespoke character, which has an almost tendon-like appearance.

Chapter Three speculated on whether an architectural design's physical integrity (as an object) is dependent on how well its material order is visually integrated with its overall building-object form (see Figures 3.16-3.18). In relation to this, both *Tendinous Bridge* and *AirBaltic's* aesthetic characteristics involve a high-resolution granular material expression that is central to their formal expression. The scale at which these operate, however, determines the degree to which they are unified with the architectural object as a whole. In *AirBaltic*, material organisation is determined at a very small scale relative to the overall roof (Figure 5.17). While it varies in depth by several metres in some areas, the overall visual effect on the large-scale building is primarily a surface texture, akin to sand settlement patterns on larger-scale dune formations. Although heterogeneous and coherent in order, there is a disconnect between material expression and the object as a whole. In contrast, *Tendinous Bridge* has a material grain that is large enough to visually impact the overall design expression, yet small enough to provide a granular resolution that varies throughout the bridge (Figure 5.20). While this may not always be achievable, such inseparable relationships between formal and material orders arguably create more object-integrity within architectural designs.

The semi-autonomous behavioural approaches to design formation explored in this book offer various means to explore these exciting poly-scalar aesthetic conditions.

Non-Optimal, Negotiated Structural Formation

The projects presented in this chapter were developed through formative self-organisational processes, where design solutions incorporated structural design criteria within a more extensive behaviour-based design methodology. This was achieved through the embedding of structural principles within a multi-agent simulation model. In some projects, structural analysis data was incorporated as an initial input, or as iterative feedback in-the-loop to inform subsequent agent-based decisions. In all cases, autonomous agent-based decisions were encoded with both design and structural criteria, and resulted in design solutions that integrated structural considerations. Although these designs are not necessarily structurally optimal, they are negotiated design solutions that attempt to resolve several conflicting design criteria, including structural considerations. The degree to which these autonomous processes operate within the design process also varies. For example, a multi-agent method is applied to the entire project in *Tendinous Bridge* but operates only on specific elements in *Fibrous Tower* or *AirBaltic*. Substantial explicit modelling and other architectural planning considerations operated alongside these autonomous processes. While such intuitive activities operating in parallel might be offensive to a process-driven purist, the narrow application of autonomous methodologies for specific tasks in these projects ensures that multi-agent methods are imbued with specific design intent[7]. In subsequent chapters, regardless of the scale and extent to which a generative process operates, it is always focused on resolving a specific spatial problem. For this reason, the multi-agent computational processes described in this book can be considered semi-autonomous in relation to the levels of autonomy discussed in Chapter One.

▼
5.21 Tendinous Bridge
a) Funicular tailoring strategy to generate geometry. (b-d) Structural deflection analysis of initial design iterations. Improvements demonstrated shifting from an open plane to a tubular surface. Highest deflection and need for reinforcement where the tube opens and transitions to flat surface.

a

b   c   d

▲

5.22 Tendinous Bridge
3D chunk study. The
unravelling of the volume
aims to de-materialise
vertical support.

Material and Fabrication Considerations

Although the projects presented in this chapter incorporate a structural strategy, their formal and topological complexity makes them difficult to realise using conventional building techniques. Most of the projects presented relied on industry approaches to casting materials off-site or on-site in moulds. *Fibrous Tower* and *AirBaltic* were designed as concrete structures, while the *Sao Paulo* and *Tendinous Bridge* employed fibre composites. *Fibrous Tower's* in-situ cast concrete design utilised relatively simple formwork consisting of interlocking void-formers positioned on two opposing sides (Figure 5.14), while *AirBaltic* used more elaborate precast concrete elements as permanent formwork for in-situ cast concrete. The economic competitiveness of the designs varied depending on the scale, location, and fabrication approach of each project. However, regardless of these factors, their manufacturing approaches were wasteful due to their heavy reliance on material formwork and labour. Several of the chapters that follow aim to address these concerns by incorporating manufacturing and construction logistics into the design approach, also employing robot systems. The next chapter demonstrates how similar approaches to multi-agent design can be developed for additively manufactured building applications while attempting to reduce material, and in some applications, enlisting robot behaviour and material properties for additional means of design expression.

Although geometric approaches to structural design mentioned at the outset of this chapter can produce exceptional architectural designs, they do not provide a means to creatively engage with robotic or material behaviours. A geometric design approach would require a separate process to address these activities, that could not feedback to inform the initial geometric design, making it difficult to develop a poly-scalar interplay between formal and material orders that could impact a design's object-integrity, as mentioned earlier. In contrast, the algorithmic approaches presented in this chapter were developed through a programmed agency that can be creatively extended into non-geometric activities such as robot manufacturing tasks. This expands the scope of design into new territories. The differences between these two approaches may not be immediately apparent in this chapter.

However, several projects presented in later chapters demonstrate the merits of behavioural approaches to design that are capable of real-time adaption to materials, environments and several other dynamic considerations that might be engaged within the conception, fabrication or construction of architecture or its constituent parts. The semi-autonomous approaches to design operating in 3D graphics media in this and the previous chapter are extended into a more extensive set of design, fabrication and construction processes in subsequent chapters, commencing with computational, robotic and material considerations for additive manufacturing in the next chapter.

Notes

1. *Thompson describes how bone-to-flesh ratios increase dramatically scaling up from a mouse to a human, or elephant (D. A. W. Thompson 1992, 20).*

2. *Although the US has a regulatory framework to allow pedestrian bridges to be built with fibre-composites, at the time in Brazil, the client did not have the appetite to pursue a non-standard approval process.*

3. *Springs are based on Hooke's Law, where a spring has a rest length, damping and stiffness value and is impacted by agents with mass and a motion vector attached to each end of each spring.*

4. *As described in Chapter Four, an agent's range of vision allows an agent to calculate its own actions based only in consideration of other entities within a distance threshold (radius) of the agent, and also only within an angle cone of vision orientated to the agent's velocity vector (field-of-view). These attributes were inspired by Craig Reynold's Boids research (Reynolds 1987).*

5. *An isosurface is used to make a volumetric mesh interpolation of a 3D point cloud.*

6. *Similar to the NAMOC project presented in Chapter Four (see Figure 4.9c).*

7. *A tight relationship must be balanced between spatial specificity and the abstraction inherent in bottom-up programmed agent behaviour. Overly specifying behaviours or spatial constraints prevents self-organised design solutions from emerging, whilst, under-specifying spatial aims provide abstract agent behaviours with insufficient instruction to deliver on architectural intent.*

▼

5.23 Tendinous Bridge
3D chunk study seen from above.

Project Credits

Habitation Sur Mars (HSM) (2004): *Kokkugia. Design Directors: Jonathan Podborsek, Roland Snooks, Robert Stuart-Smith.*

Sao Paulo Bridge (2012): *Robert Stuart-Smith Design: Design Director: Robert Stuart-Smith. Design Team: Gilles Retsin, Daniela Pais-Hernandes.*

Taipei Performing Arts Center (2008): *Kokkugia. Design Directors: Roland Snooks & Robert Stuart-Smith. Design Team: Brad Rothenberg, Elliot White, Matt Howard.*

Fibrous Tower (2008): *Kokkugia. Design Directors: Roland Snooks & Robert Stuart-Smith. Design Team: Timo Carl, Juan de Marco.*

AirBaltic Terminal (2010): *Kokkugia. Design Directors: Roland Snooks & Robert Stuart-Smith. Design Team: Casey Rehm, Matt Choot, Edwin Liu.*
Buro Happold LA + London. Design Manager: Steve Chucovich, Director of Aviation: Alan Regan, Structural Engineering: Matthew Melnyk.

Tendinous Bridge (2020): *Robert Stuart-Smith Design: Design Director: Robert Stuart-Smith. Design Team: Patrick Danahy.*

Image Credits

Figures 5.2-5.4: *Kokkugia Ltd, Robert Stuart-Smith, Roland Snooks, Jonathan Podborsek*

Figures 5.1, 5.8-5.16: *Kokkugia Ltd, Robert Stuart-Smith and Roland Snooks*

Figures 5.5-5.7, 5.17-5.22: *Robert Stuart-Smith Design Ltd, Robert Stuart-Smith*

References

Burry, Mark. 2013. "From Descriptive Geometry to Smartgeometry: First Steps Towards Digital Architecture." In Inside Smartgeometry: Expanding the Architectural Possibilities of Computational Design. Vol. 9781118522479. doi:10.1002/9781118653074.ch13.

Chilton, J, and H Isler. 2000. Heinz Isler: The Engineer's Contribution to Contemporary Architecture. Engineer's Contribution to Architecture Series. Thomas Telford.

Cuito, A, and C Montes. 2003. Gaudi: Complete Works. Cologne: DuMont.

Gargiani, R, A Bologna, and J Haydock. 2016. The Rhetoric of Pier Luigi Nervi: Concrete and Ferrocement Forms. Treatise on Concrete. CRC Press LLC.

Garlock, M E M, and D P Billington. 2008. Félix Candela: Engineer, Builder, Structural Artist. Princeton University Art Museum. Princeton University Art Museum.

Graefe, Rainer. 2020. "The Catenary and the Line of Thrust as a Means for Shaping Arches and Vaults." In PHYSICAL MODELS, 79–126. doi:https://doi.org/10.1002/9783433609613.ch3.

Hooke, R. 1676. A Description of Helioscopes and Some Other Instruments. Cutlerian Lectures. T.R.

Isler, Heinz. 1980. "Structural Beauty of Shells." IABSE Congress Report 11. IABSE: 147. doi:10.5169/seals-11239.

Kilian, Axel, and John Ochsendorf. 2005. "Particle-Spring Systems for Structural Form Finding." Journal of the International Association for Shell and Spatial Structures 46 (August): 77–84.

Nejur, Andrei, and Masoud Akbarzadeh. 2021. "PolyFrame, Efficient Computation for 3D Graphic Statics." CAD Computer Aided Design 134. doi:10.1016/j.cad.2021.103003.

Pedreschi, R. 2000. The Engineer's Contribution to Contemporary Architecture: Eladio Dieste. Engineer's Contribution to Contemporary Architecture. Thomas Telford.

Pigram, Dave, and Wes McGee. 2011. "Formation Embedded Design: A Methodology for the Integration of Fabrication Constraints into Architectural Design." In Integration Through Computation - Proceedings of the 31st Annual Conference of the Association for Computer Aided Design in Architecture, ACADIA 2011. doi:10.52842/conf.acadia.2011.122.

Reynolds, Craig W. 1987. "Flocks, Herds, and Schools: A Distributed Behavioral Model." In *Proceedings of the 14th Annual Conference on Computer Graphics and Interactive Techniques, SIGGRAPH 1987*, 25–34. Association for Computing Machinery, Inc. doi:10.1145/37401.37406.

Rippmann, Matthias, and Philippe Block. 2013. "Funicular Shell Design Exploration." In *ACADIA 2013: Adaptive Architecture - Proceedings of the 33rd Annual Conference of the Association for Computer Aided Design in Architecture.*

Schanz, S. 1995. *Frei Otto, Bodo Rasch: Finding Form : Towards an Architecture of the Minimal.* Axel Menges.

Stuart-Smith, Robert. 2011. "Formation and Polyvalence: The Self-Organisation of Architectural Matter." In *Ambience'11 Proceedings*, edited by Annika Hellström, Hanna Landin, and Lars Hallnäs, 20–29. Boras: University of Borås.

Thompson, D A W. 1992. *On Growth and Form.* Dover Books on Biology Series. Dover Publications.

Thompson, D'arcy W. 1915. "Galileo and the Principle of Similitude." *Nature* 95 (2381): 426–27. doi:10.1038/095426a0.

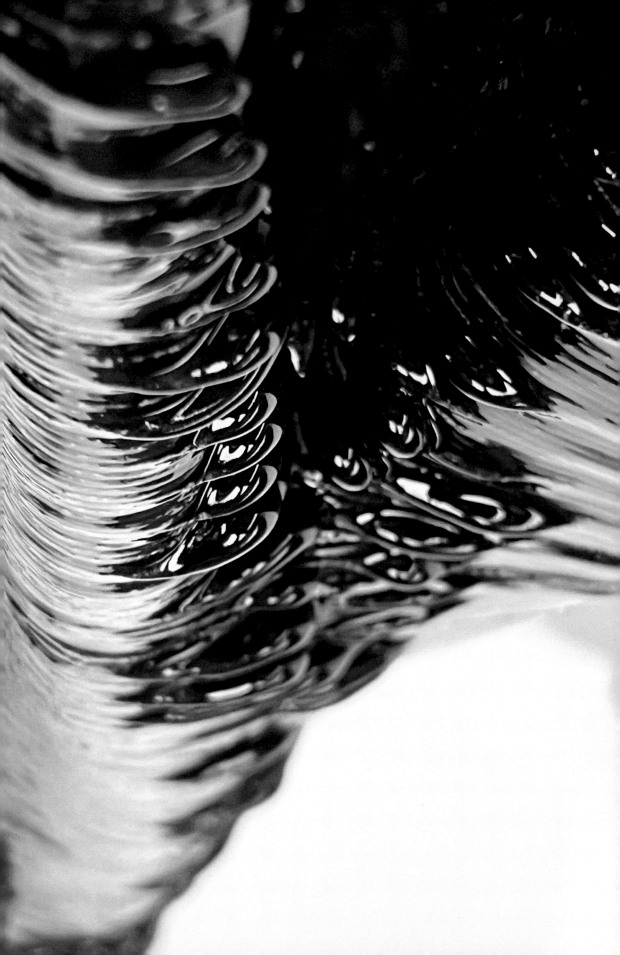

6 | Additively Manufactured Architecture
Form, Topology and Material organisation

In the previous chapters, generative architectural design approaches were presented that addressed diverse design criteria and produced a distinctive aesthetic character that was intrinsic to each project's custom-developed algorithm. This chapter expands this work towards the design of an additively manufactured architecture, introducing additional design considerations. While the possibilities of additively manufactured architecture will continue to broaden with continued technological advancements, this chapter focuses on some core principles that can be addressed using current manufacturing infrastructure. A behavioural methodology is first established to integrate approaches to formation, topology, and structural efficiency. An agent-based approach to material placement and robot manufacturing is then explored, synthesising the previous three chapters' approaches to computation and production. Material-physics simulation modelling also provides visualisation and analysis capabilities to support this design workflow. Before discussing this work, a context must be established around the transformative potentials of additive manufacturing, the building-scale manufacturing technologies available today, and the possibilities these afford architectural design. A reader with significant knowledge in this area might wish to skip this section and move ahead a few pages to *"An Initial Set of AM Architectural Design Principles"*.

Game-Changing Manufacturing

◀◀

6.1 Viscera/L
Design for an additively manufactured pavilion developed using topoForm#2 methodology

◀

6.2 matForm
Detail photograph from an additively manufactured ceramic column geometry

Additive Manufacturing (AM), also known as "3D printing" or "Rapid Prototyping," is a method that joins material in layers to build objects from 3D model data (ISO/ASTM 2021, 1; Gibson, Rosen, and Stucker 2015, 1). AM is a rapidly growing global market, currently valued at USD $13.4 billion, with potential for near-term growth up to USD $550 billion (Bromberger and Kelly 2017; Bromberger, Ilg, and Miranda 2022). It is transforming the design and manufacture of products and is increasingly used in the construction industry for off-site prefabrication and on-site construction applications. Examples of AM in construction include University of Loughborough's pioneering research into AM building parts (Buswell et al. 2007), the multi-storey prefabricated buildings of Winsun Decoration and Engineering Co™ in China (Sevenson 2015), and US firm Icon™'s on-site constructed suburban houses or their collaboration with NASA™ to develop lunar construction capabilities (Dreith 2022). AM buildings can be realised in a fraction of the time, cost or environmental impact compared to traditional methods. For example,

SQ4D recently constructed a 1900 square-foot home using less than $6000 worth of cementitious material in less than 48 hours[1] (Essop 2020). The potential long-term benefits of additively manufactured architecture are therefore significant and include reshaping the way buildings are designed, manufactured, used, demolished, recycled, developed, managed, and sold. This impact is perhaps best summarised in the following opportunities:

1. Design-agnostic economics
AM operates from software-generated instructions derived from a 3D model involving minimal supervision or specialised labour costs. Complex geometric designs are not discriminated against as material volume and time primarily drive cost.

2. One-off manufacturing every time
Unlike most formative manufacturing[2] that requires moulds or dies that are only economically competitive with repetitive use, each AM part can be geometrically unique, enabling economically competitive small-batch production and bespoke architectural design, repair, replacement, or remanufacturing of unique parts on demand (Ford and Despeisse 2016, 1581).

3. Increased design complexity and efficiency
Benefits (1) and (2) provide greater aesthetic freedom while more complexity in form and topology can also enhance a part's structural efficiency and reduce material usage, resulting in significant economic and environmental savings. By varying architectural geometry within or across parts in relation to structural forces, exciting opportunities abound to explore relationships between space, form, structure, and ornamentation.

4. Reduced parts/assembly requirements
Benefits 1-3 allow multiple individual parts to be fabricated as a single object. Additionally, many AM technologies can fabricate nested parts, something difficult to achieve with other manufacturing methods (Lipson and Kurman 2013, p.20, Location 451). For example, Relativity Rocket's Terran R™ reusable rocket uses less than 1% of the parts of a traditional rocket ("Relativity Space" 2022). If this can be done with spacecraft, it can be done with buildings. Increasing the complexity of individual parts could simplify building assembly and create exciting opportunities for custom architectural components that incorporate additional functionalities or effects.

5. Complex material composites
Individual material grading and the differentiated organisation of multiple materials can be achieved in a single AM part (Garcia et al. 2018; Oxman, Keating, and Tsai 2012). Scientists Lipson and Kurman claim that future AM technologies will also allow designers to go beyond form to design behaviour, such as self-healing capabilities, 3D electrical circuits, and individualised medical performances (Lipson and Kurman 2013, p.15, Location 346-379). Researchers are already demonstrating 4D designs that transform over time (Tibbits 2017). These AM composites could challenge established boundaries between buildings, people, objects, and the environment, leading to new social, functional, experiential, or aesthetic possibilities.

6. Reduced material waste and environmental impact
In aerospace, complex parts are often made by subtractive manufacturing processes that cut away excess material from raw material blocks four

to twenty times the volume of the final part (Ford and Despeisse 2016, 1577). This waste cannot be easily re-used. With additive manufacturing, there is almost no waste because the process builds up only the volume of the part. Powdered AM material methods also often use recycled metals or plastics, with 95-98% of any remaining material used to make other parts (Ford and Despeisse 2016, 1579).

7. Immediacy

Most architectural designs must be finalised well in advance of manufacture. Lead times for securing raw materials, producing moulds, scheduling labour, equipment and space or post-processing parts can take months to years. Except for raw material supply logistics, AM offers a direct and expedited path from design to fabrication with minimal impedance. A design could even continue to be modified throughout manufacturing providing already fabricated parts are not re-designed. There are also possibilities for near-instant architecture that might offer a rapid response to temporal concerns or opportunities.

8. Adaptive, distributed manufacturing

Given benefit (7), AM supply chain logistics are simplified compared to other manufacturing methods, allowing smaller, distributed production facilities (Gibson et al. 2021, 9). Decentralised manufacturing can support reduced production times and transportation costs, and local sourcing of materials where possible (Ford and Despeisse 2016, 1583). Coupled with benefit (7), economies of scale will have less influence, with project feasibility less dependent on the sise of the project or production facility, as fabrication can be distributed across one or several factories with little cost overhead.

9. Opportunities for remanufacturing and replacement parts

AM enables faster and cheaper replacement parts than traditional manufacturing. Remanufactured parts use a fraction of new material compared to new parts, while in some instances, new AM replacement parts can be produced up to ten times faster and at a lower cost than stockpiled parts (Ford and Despeisse 2016, 1579–80; Caterpillar 2022).

10. New business models and user experiences

As discussed in Chapter Two, AM is a Fourth Industrial Revolution technology that leverages cyber-physical systems to allow for more user-customisation. Leveraging 1-9 above, there is ample opportunity for designers to develop unique and inspirational architectural experiences. Collaborative teams of architects, fabricators, developers, technologists, and others may also explore hybrid business models to further expand creative possibilities.

The work presented in this chapter is most aligned with benefits 1-4. The importance of this chapter's developments, however, is contextualised within this larger, emerging body of possibilities.

Additive Manufacturing Technologies

Since Chuck Hull's 1986 patent for a stereolithographic AM process that projected UV light onto UV curable resin (Hull 1986), a wide range of AM methods and materials have been developed. The American Society for Testing and Materials (ASTM) has defined a standard terminology that consolidates these into seven umbrella categories. Compared to the potential long-term AM building possibilities just mentioned, four of ASTM's categories already demonstrate exciting applications at the scale of buildings or building parts:

a

b

c

◀

6.3 CiPD
Centre for Innovation
& Precision Dentistry,
University of
Pennsylvania: a) new
building floats above
the existing courtyard,
bridging between three
existing buildings,
b-c) pilotis transfer
structure with few
touch-down points
elevates the building
above the courtyard.
The reflective surface
is planned to be WAAM
permanent formwork or
polished cast-in-place
concrete using MEX-AM
sacrificial formwork.

- *Binder Jetting (BJT):* a liquid binder fluid is jetted into a bed of powdered PMMA (Voxeljet), plaster (3D Systems) or other material (Gibson, Rosen, and Stucker 2015, 239). BJT has been used for final parts, moulds for precast concrete slab designs, and for exquisite interior installations (D-Shape Enterprises 2022; Lowke et al. 2018; Hansmeyer and Dillenburger 2013; Jipa, Bernhard, Meibodi, et al. 2016). The commercial viability of BJT methods for building-scale final parts or reusable moulds is uncertain due to their high cost and low durability. However, advancements in wood BJT using recycled materials may help overcome these challenges (Forust 2021).

- *Directed Energy Deposition (DED):* wire arc additive manufacturing (WAAM) is a DED method that deposits metal in layers using a robot arm with standard welding equipment. WAAM parts are typically machine-finished to remove rough exterior surfaces. WAAM has shown a 60% reduction in fabrication time and 15-20% in post-machining time compared to subtractive manufacturing of large metal parts (Gibson et al. 2021, 304). While these benefits are not as significant for architectural applications that involve less machining and more welding of standardised parts, the ability of WAAM to support structurally efficient geometries suggests great potential for saving material and labour. Examples such as a canal bridge in Amsterdam by MX3D™ and Relativity Rocket™'s reusable suborbital rocket, Terran R™, demonstrate WAAM's feasibility for building-scale applications ("Relativity Space" 2022; MX3D 2020).

- *Powder Bed Fusion (PBF):* one method, selective laser sintering (SLS) incrementally builds a bed of metal powder (such as titanium, steel, or aluminium) in layers, fusing the metal using a laser (Gibson et al. 2021, 126,167). Post machine-finishing, metal PBF processes offer outstanding manufacturing precision and quality, and are already cost-competitive for small-batch production in aerospace and medical applications. While currently not cost-effective for large-scale architectural work, the manufacturing time and cost of PBF methods are expected to decrease significantly in the future, making it a potential option for high-strength or intricate architectural parts (Gibson et al. 2021, 167).

- *Material Extrusion (MEX):* encompasses various techniques such as fused deposition modelling (FDM), chemical reaction bonding (CRB), and liquid deposition modelling (LDM). These techniques involve the continuous extrusion of material in layers that bond with previous layers. FDM commonly uses plastics like ABS, PLA, and PC, and can include carbon fibre, Kevlar, or fibreglass to create composite materials. LDM uses materials like clay, wood, organic fibre composites, and hydrogels. CRB/C/Concrete extrudes cementitious paste that cures after deposition (Gibson et al. 2021, 190,196; Wohlers and Gornet 2015). FDM has been used for precast and in-situ concrete formwork, lattice-structured pavilions, and furniture (Tang and Shroyer 2020; Martin 2021; Soler, Retsin, and Garcia 2017). CRB/C/Concrete is in widespread use for the in-situ construction and off-site prefabrication of concrete buildings. Notable examples include Icon (Dreith 2022), Apis Khor (Apis Cor 2017), Xtree3D (XtreeE 2022) and others, together with research by institutions including Loughborough (Buswell et al. 2007), USC (Khoshnevis et al. 2001), Eindhoven (Parkes 2021) and ETHZ (Anton et al. 2020), amongst others.

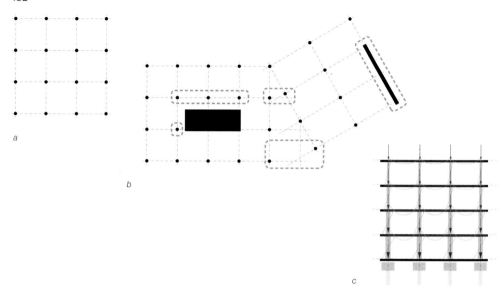

a

b

c

▲

6.4 Anisotropy
While building structures
are idealised as
isotropic, force loads
are asymmetrical.
This results in built
matter embodying an
anisotropic distribution
of forces. a) plan of
an ideal standardised
column-grid, b) real-
world site constraints
can lead to a non-
standard structural
grid, c) section of
asymmetrical loading in
a multi-storey building.

Whilst other ASTM categories offer novel capabilities, the methods highlighted above are already gaining traction within building and manufacturing industries, and support fabrication in metals, concrete, ceramics, woods, and plastics. They are also capable of fabricating complex geometric designs that can support greater material and structural efficiencies. This chapter proposes a conceptual set of design principles and a custom set of generative design methodologies that could leverage the capabilities of these already accessible technologies.

An Initial Set of AM Architectural Design Principles

Architects who design AM buildings face a clear ethical problem – how best to explore aesthetic design expression to leverage newfound manufacturing freedoms while ensuring a design has sufficiently optimised material volume, which has both an economical and environmental impact. The research outlined in this chapter addresses this dilemma, exploring aesthetic design expression in relation to volumetric (material) and structural considerations. The approach is primarily explored within generative algorithmic design (more so than manufacturing) and emphasises four key AM architectural design concepts: multifariousness, anisotropy, topology, and material effects. These are addressed through a behaviour-based approach to topological and material formation.

A Multifarious, Integrated Whole

Classical architecture, as described by Vitruvius (Pollio and Morgan 1960), sought to create architectural order through proportional relationships between building elements. Modernist architect Le Corbusier's Five Points of Architecture also emphasised flexible design opportunities that arise from the conceptual separation of elements such as floors, column grids and façades (Corbusier 2013). However, in Baroque and Art Nouveau periods, architects such as Francesco Boromini (Wölfflin, Murray, and Simon 1966), Victor Horta (Borsi and Portoghesi 1996) or Hector Guimard (Guimard et al. 1992) demonstrated through painterly, compositional, or formal methods, an alternative means of architectural expression that operates as a complex heterogeneous whole. Although AM is often associated with monolithic designs, it can be used for the prefabrication of multi-part assemblages or combined with parts manufactured by other means and materials. While most built AM architectural designs are relatively conventional, AM product designs for shoes (Finney 2023), bicycles (Scott 2016; Takahara 2014), furniture (Boruslawski 2016), or mechanical parts ("Relativity Space" 2022) demonstrate that intricacy and formal complexity can be strategised to provide lightweight, materially efficient designs that are aesthetically diverse. The potential of AM architecture also lies in both material-scale and object-scale design thinking, challenging the traditional principles of modernist architects like Hannes Meyer that asserted buildings are simply an assemblage of available products or parts (Meyer 1976). Architecture can go beyond being an assemblage of known elements or a monolithic whole devoid of architectural elements and embrace the interrelation of diverse parts that vary in scale, order, purpose, or kind, recasting architecture as a multifarious whole. Cognisant that most architecture is comprised of several parts, the work presented in this chapter aims to support the development of multifarious architectural designs.

Anisotropy
The material efficiency of AM architecture is correlated to its structural efficiency. Unfortunately, buildings are rarely optimal structures. Many building sites introduce additional complexities to a design problem that make it challenging to implement ideal structural geometries. Architects must also resolve several conflicting design criteria within each project such as planning, user, or site-related constraints, which can take precedence over structural or geometric concerns. These can lead to non-standard or asymmetrical distributions of structural elements within a building (Figure 6.4). For example, a schematic design for the *Center for Innovation and Precision Dentistry (CiPD)* proposes a pilotis structure to elevate a new building above an existing courtyard (Figure 6.3). To safeguard the existing public space, inclined columns are placed in specific positions within the courtyard to connect the ground to a larger array of columns above (the ceiling operates as a transfer structure), creating an anisotropic structural condition. In *CiPD*, architectural concerns clearly trump structural efficiency concerns. Privileging structural over architectural criteria would involve jettisoning the courtyard to allow the new building to sit directly on the ground.

▼
6.5 Quantitative Metrics
Aesthetics must be evaluated relative to: a) material volume, b) surface area and c) structural performance.

While many buildings may appear isotropic, force distributions within them are often anisotropic. Spatiotemporal live loads, such as those from people, furniture, or snow, are distributed heterogeneously. Even regular structural systems, like the array of columns and slabs in a multi-

a   b   c

Additively Manufactured Architecture

6.6 Viscera/L
*Design for an additively
manufactured
pavilion developed
using topoForm#2
methodology. Close-up
of v-column and its
transition to ceiling.*

storey building, exhibit heterogeneous dead-load force distributions. The loading increases from top to bottom as lower-level columns support the weight of the floors above. Horizontally, most floor systems abstract structural forces to either a one-axis or two-axis loading condition, requiring forces to take an indirect path to vertical support locations. An early response to this can be seen in Engineer Pierre Luigi Nervi's application of force-driven design principles in his 1949 isostatic slab patent, and its use in the Manifattura Tabacchi Complex design and other exceptional projects (Gargiani, Bologna, and Haydock 2016). However, Nervi's designs were limited by the time and cost constraints of cast-in-place concrete manufacturing methods, which necessitated repetitive use of moulds, limiting geometric variation. In contrast, present-day AM methods can accommodate greater formal and material complexity, allowing for topological approaches to design and engineering that are both intricate and efficient. This enables fine-tuning of a building's material construction to closely relate to variations in the structural forces throughout different parts of the building.

Topology
The four AM technologies highlighted earlier differ in material and manufacturing constraints, but all support substantial formal and topological complexity. Design-engineering approaches can leverage these capabilities to improve material and structural efficiency. One commonly used method, Topological Structural Optimisation (TSO), removes unnecessary material from a 3D design in areas with low structural utilisation, resulting in a more efficient and higher-genus topological geometry for a given structural loading condition (Eschenauer and Olhoff 2001). TSO has been used in the design of novel building floor slabs (Jipa, Bernhard, Dillenburger, et al. 2016), pavilion canopies (Louth et al. 2017), and large-span halls (Sasaki, Ito, and Isozaki 2007). However, TSO is not versatile in logic or aesthetics. It tends to override or over-constrain a design's formal character depending on the design stage and scale in which it is implemented, potentially undermining design-authorship (e.g., to manipulate a design's overall formal silhouette or at a smaller scale to perforate a design model). For greater aesthetic control, one can 3D model a multi-genus topological design explicitly or procedurally; however, these methods limit adaptation to structural analysis data while explicitly modelling such complexity is also arduous. In contrast, algorithmic approaches to generative topological design can incorporate structural and geometric efficiencies while embodying a designer's approach to formal character.

Material Effects
In MEX-AM methods, materials phase change from a liquid to a solid during a continuous extrusion process, opening up the possibility of variable material characteristics being registered in a design outcome. Such variation can also impact interlayer bonding, essentially grading structural integrity and material density. In recognition of this, an integrated approach to the production of material effects and their high-resolution structural analysis is explored in this chapter.

Behaviour-Based Approach to Material Formation
Although AM architecture is relatively nascent, computational design approaches have been developed for architectural elements like columns and floor slabs (Anton et al. 2020; Aghaei Meibodi et al. 2017) or bridges (ETHZ BRG and Zaha Hadid ZHACODE 2021). To date, these approaches have not extensively addressed the variable

136

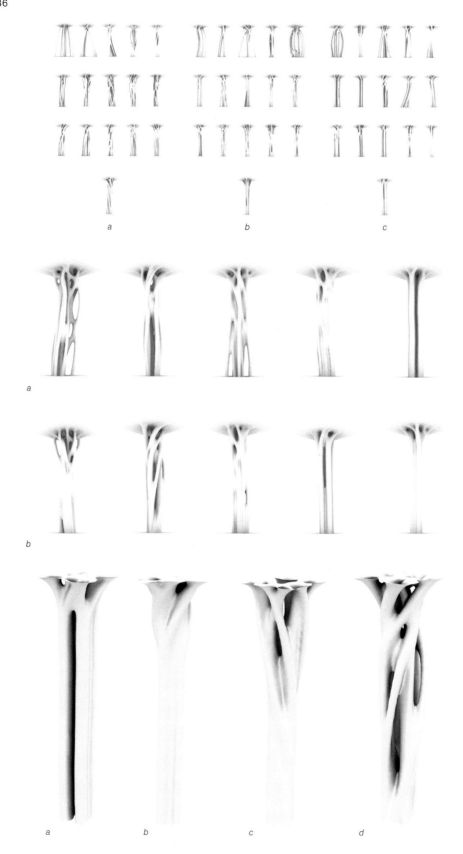

a   *b*   *c*

a

b

a   *b*   *c*   *d*

◀

6.7 TopoForm#1
Abbreviated sample of three evolutionary runs (a-c) of a column-like geometry. Initial generation at top does not produce feasible columns. These emerge during later generations of evolution towards the bottom.

◀

6.8 TopoForm#1
A selection of column-like geometries: a) produced mid-way through an evolutionary process, b) at end of evolutionary process.

◀

6.9 TopoForm#1
Four column results (a-d) from different evolutionary sequences, each produced from variations in the weighting of several fitness criteria.

▼

6.10 TopoForm#1
Fitness criteria used in the evolutionary solver proportionally weights the following metrics obtained from geometric and structural analysis: a) structural deflection and stress (b), c) surface area, d) volume, e) curvature, f) horizontal sectional area at 3 intermittent heights, g) base footprint area.

spatial or structural conditions found in buildings, nor been employed in non-spatial design activities addressed in this book such as robot manufacturing or multi-robot construction logistics. Beyond addressing architectural and structural spatial design criteria, the agent-based behavioural design approach advocated for in this book can also be used in robotics and material applications, expanding design activities into production processes. In this chapter, agent-based methods are used to design multi-genus topological geometries for AM with corresponding material effects produced through robot programming. Design is recast as a topologically and materially formative process, capable of adapting to diverse spatial and structural conditions.

Generative Approaches to Anisotropic Topological Design

Two research projects, *topoForm#1* and *#2*, explore design approaches for an additively manufactured architecture. Both involved the development of a custom multi-agent design algorithm similar to those employed in previous chapters (Stuart-Smith 2011), coupled with geometric and structural analysis methods. *topoForm#1* focuses on the topological formation of column-like geometries, while *topoForm#2* extends the approach into more flexible architectural scenarios, adapting to diverse spatial and structural arrangements. Design and analysis methods used in both approaches are materially agnostic and applicable to the four ASTM AM technology categories mentioned earlier. Although concrete was employed as a proxy material for structural analysis purposes, the methods can be further calibrated in future work to address material-specific performance requirements. In both projects, multi-agent simulated motion trajectory curves are interpolated as a volumetric mesh, to create a hollow-core surface geometry.

topoForm#1
topoForm#1 generates topological design solutions for column-like elements while optimising for several structural and geometric conditions using an evolutionary process (Stuart-Smith, Danahy, and Rotta 2020). An evolutionary solver tests several design solutions and cross-breeds input parameter values from more successful outcomes to produce better-performing results over time (generations). Evolutionary algorithms, first introduced by John Holland in 1975 (Holland 1975), have been widely used for optimisation tasks (Mitchell 1998). *topoForm#1*, however, inspired by John Frazer's Etruscan Column

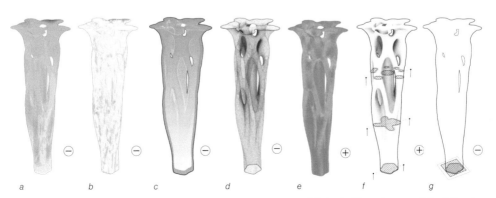

a   b   c   d   e   f   g

Additively Manufactured Architecture

a

b

c

project (Frazer 1995) and Richard Dawkins' Blind-Watchmaker algorithm (Dawkins 1986), employs an evolutionary solver within a generative design process. It mixes bottom-up agent-based behavioural rules with top-down geometric and structural analysis, to evolve agent behaviour towards higher-performance results, while also developing an aesthetic character that emerges from its local agent-based interactions (Figures 6.7 – 6.11).

A variable number of agents are seeded along three concentric rings located on the x-y axis origin, 3.5m high in the z-axis. Agents create a column-like bundle of curves as they descend to zero altitude over time (Figure 6.11). Each agent's motion is governed by steering behaviours (Reynolds 1999) that involve seeking, avoiding, and aligning with other agents[3], as well as moving towards the 3D environment's ground plane. Behaviours calculate vectors that are proportionally weighted (scaled), summed and limited to a maximum velocity. These methods are also influenced by variables like radius, which limits behaviours to interactions with nearby agents, and field of view, which restricts agent perception to a specific angle cone of vision (see Figure 4.5). The simulation captures a self-organised order in the agents' motion trajectory curves, which are then interpolated into an isosurface volumetric mesh. All simulation variables were incorporated into an evolutionary solver's[4] gene pool to allow for the evolution of both the spatial set-up and agent behaviour. In each generation of the evolutionary sequence, design outcomes were structurally and geometrically analysed for column-related performance criteria. Metrics obtained were combined into a single fitness value that aimed to reduce material volume, improve structural performance, and minimise footprint area (Figure 6.10).

A single evolutionary simulation produced a population of thirty column-like structures for each generation. Over several hundred generations, a matrix of design options was created which became increasingly effective at meeting the defined fitness criteria. *topoForm#1* was used in multiple evolutionary sequences, each with different weights assigned to the fitness criteria. The results were exciting in many ways: designs were diverse both topologically and geometrically and had a unique character derived from their agent-based behavioural formation.

◀

Evolved designs significantly improved upon initial designs in the evolutionary sequence and embodied several design and structural considerations, such as avoiding excessive overhangs that could lead to collapse during AM MEX manufacturing. The method demonstrates a successful hybridisation of bottom-up multi-agent behaviour and top-down evaluation procedures. Designer intuition also played a key role in developing the multi-agent algorithm, its spatial set-up, and defining the a priori range of variation in agent behaviours exposed to evolutionary variance. Although the method was calibrated for low-resolution AM, an intermediate scale of complexity has been explored in sacrificial AM formwork for cast metal parts (Figure 6.12). While *topoForm#1* could potentially be scaled to utilise larger populations of agents for higher-resolution designs, the method is unable to address larger-scale architectural orders, as it incorporated no means to respond to the spatial and structural complexities of variable architectural conditions.

topoForm#2

topoForm#2 explores the design of geometrically integrated ceilings, slabs, columns and walls, adapting to different planning and structural constraints. Developed and tested at the scale of one structural bay, *topoForm#2* generates anisotropic topological designs (Figure 6.13). Unlike *topoForm#1*, which seeded agents in specific locations, *topoForm#2* adapts to diverse spatial and structural conditions, populating agents across a ceiling surface with varying density, location and orientation in response to structural analysis data and support locations[5] (Figure 6.14). Agent steering behaviours respond to nearby agents, agent's motion trails, structural data, and vertical support conditions (Figure 6.13). Agents operate under two states, each of which weighs these behaviours differently. State "1" is used from the outset to navigate across the ceiling surface. When agents arrive within range of a vertical support, they switch to state "2", which prioritises seeking and aligning to the support over other behavioural influences. Like *topoForm#1*, agents' motion trajectories are interpolated as a volumetric mesh, except in *topoForm#2*, each trajectory's control points adjust their depth relative to the ceiling based on local agent density and structural performance values, and proximity to support locations.

▶

a

b

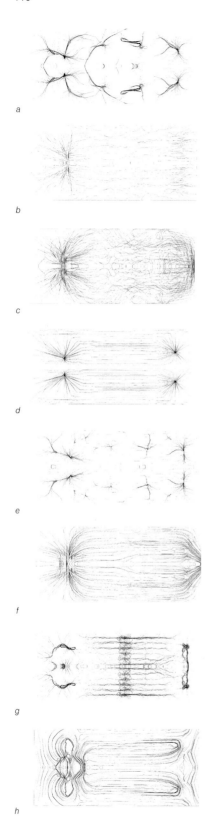

a

b

c

d

e

f

g

h

While no optimisation routines were used in *topoForm#2*, analytical methods were developed to support the integration of an evolutionary or reinforcement learning optimisation process. Design outcomes were geometrically analysed for surface area, volume, structural deflection and principal stress, alongside embodied carbon estimates and visual character analysis methods (discussed in the next chapter). A matrix of design outcomes was created to test *topoForm#2's* adaptability to different spatial and structural arrangements and evaluated using the mentioned performance metrics (Figure 6.15). topoForm#2 demonstrates a generative approach to anisotropic topological design for AM architectural applications that is adaptive to various diverse spatial and structural conditions. *topoForm#2* is also being used to design *Viscera/L*, a pavilion project currently under development that also incorporates additional fabrication methods discussed in Chapter Eight. Compared to traditional reinforced concrete construction, *Viscera/L* is a lean body mass without epidermal layers – a viscera of sorts. It combines ornament and structure through a maximalist approach to complexity and a minimalist approach to material volume (Figures 6.6, 6.16, 6.17).

The *topoForm#1* and *#2* design methods increase topological complexity to provide structural and material efficiencies. This was achieved through the programming of semi-autonomous agents in 3D graphical space. The approach can also be expanded into physical domains, to inform the motion of an industrial robot MEX-AM material process to produce material-scale design effects. To explore this, a generative design approach was developed to control AM deposition trajectories during a robotic manufacturing process.

Additively Manufactured Material Effects

Currently, most building-scale AM uses MEX technologies that involve the continuous extrusion of viscous materials such as concrete, ceramics or plastics. 3D designs are typically abstracted into a series of horizontal contour curves that are manufactured in one long continuous ascending extrusion. A robot system such as a 3-axis gantry or 6-axis industrial robot arm is used to move a material extruder end-effector tool with a constant extrusion flow rate along a specific tool path. As a programmable machine, the robot's tool path can embody variations in trajectory and velocity that, together with gravity, cause a viscous material to settle in

◀

6.13 TopoForm#2
*Agent motion behaviours.
a) agent: cohesion, b)
agent: separation, c)
agent: alignment, d)
attractor: seek, e) agent
trails: seek, f) agent trails:
alignment, g) structural
stress: seek, h) structural
stress: alignment.*

varied ways to produce diverse aesthetic effects. Several designers have written algorithms to explore these effects in ceramics and concrete (Rael and Fratello 2017; Anton et al. 2020). A series of concrete panels fabricated during *SituatedFabrications* – a 2016 visiting professorship fabrication studio at the University of Innsbruck (Stuart-Smith 2018) demonstrate the aesthetic impact of such material effects (Figures 6.19, 6.20).

Uniquely, in *SituatedFabrications*, material deposition is not constrained to an ascending sequence of aligned contours. Instead, multi-agent motion behaviour was used to generate a series of robot tool paths that adapted to the iterative placement of void-former foam parts throughout manufacturing, resulting in concrete panels that contain voids and a degree of porosity (discussed in Chapter Eight). The manufactured outcomes also embody substantially more intricate and expressive material effects than those described in the robot's toolpath due to the concrete's viscosity producing an additional squiggle relative to the robot's velocity (Figure 6.18). Such approaches are influenced by the design's geometry and the gravitational orientation of manufacture that together cause variations in material density, and impact the aesthetic and structural performance of a fabricated outcome. MEX-AM material effects are therefore best-strategised relative to a part's geometrical and structural properties, and its orientation during production. Unfortunately, established MEX-AM design methods were not sufficiently integrated into material simulation and structural analysis workflows, hindering the ability for designs to be modelled, analysed or visualised prior to manufacture. Previous approaches had primarily operated through scanning-in-the-loop (Im, AlOthman, and del Castillo 2018), necessitating extensive physical prototyping, and making iterative design development challenging.

A design approach to MEX-AM material-gravitational effects was conceived that is adaptive to variations in geometrical and structural conditions and provides a means to assess an outcome's aesthetic and structural impact prior to manufacture. The method involved the development of two custom software programs, one to simulate MEX-AM material behaviour, and another to design material effects and provide manufacturing instructions for industrial robot AM (Stuart-Smith, Danahy, and Rotta 2020).

▶

6.14 TopoForm#2
*Agent seeding in relation
to spatial and structural
data: a) structural
analysis of ceiling,
b) support threshold
contours, c) density of
agent setout adapts to
contours and structural
data, d) supports
operate as attractor field.*

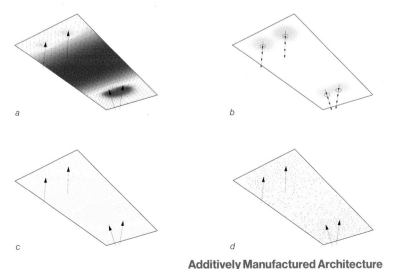

a　　　　*b*

c　　　　*d*

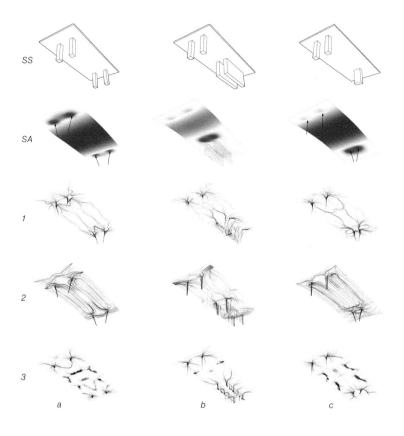

SS

SA

1

2

3

a   b   c

matSim: Material-Physics Simulation
matSim simulates the MEX-AM production process. In approximating a geometric outcome and estimating interlayer bonding, it also supports visual (Figures 6.23) and structural analysis of design outcomes prior to manufacture (Stuart-Smith, Danahy, and Rotta 2020). The software is based on particle-spring material-physics principles, commonly used in movies and computer games. In *matSim*, extruded material is abstracted to a particle-spring chain that dynamically repositions as it settles under the influence of multiple forces. A series of particles (points) with defined mass are created over time from a moving robot end-effector position. Successive particles are linked by springs modelled on Hooke's Law, taking into account stiffness, rest length, damping, velocity, and mass (Lengyel 2012, 457). The particle-spring chain settles, affected by the robot's velocity, gravity and a series of functions that approximate collisions, bonding, slippage, and collapse. While it would be possible to perform structural analysis on a volumetric mesh interpolation of the simulation, this would be time-consuming due to the high resolution of detail. Instead, a considerably faster, frame (stick-node) structural analysis is performed. To support this, connective links are created to represent bonding between layers and included in an exceptionally high-resolution structural analysis to approximate the structural integrity of a manufactured geometry (Figure 6.23c).

Observations from clay and concrete MEX-AM using an extrusion nozzle with a circular cross-section revealed that when the nozzle is positioned above a previous layer by a distance smaller than the nozzle diameter, the

◄

6.15 TopoForm#2
*Generative design
catalogue of:
(SS) three different
spatial and support
configurations (a-c),
each structurally
analysed (SA) and
tested with three
different rule sets (1-3).*

material extrudes under compression, spreading outward and resulting in a wider extrusion. Similarly, increasing velocity while maintaining the same extrusion flow rate leads to a thinner extrusion. To approximate these effects, in *matSim* a 3D mesh is generated around the springs, with cross-sections adjusted relative to robot velocity and the extruder tool's vertical offset above previously deposited layers.

By simulating material-based gravitational effects to support visual and structural analysis, *matSim* allows for extensive design exploration before production. *matSim* was developed to support a design software called *matForm*, which generates material extrusion tool paths. Together these methods offer unprecedented capabilities for design workflows that orchestrate gravitational-material effects within MEX-AM.

matForm: Agent-Based Robot Deposition
matForm is a generative design software that creates a tool path for industrial robot AM with custom material-gravitational effects relative to a user-specified 3D model input (Stuart-Smith, Danahy, and Rotta 2020). It supports both physical manufacturing and digital simulation activities. Although *matForm* is primarily used for offline programming (computing a robot tool path in advance of its execution), it was developed to also be compatible with online programming applications (real-time robot motion planning). To support this dual purpose, *matForm* extends agent-based programming research discussed earlier, ensuring its iterative decisions are made in response to locally perceived knowledge (from a virtual 3D environment, adaptation to physical environments is discussed in later chapters).

▼

6.16 Viscera/L
*Pavilion design
showcasing
asymmetrical column
configuration and
anisotropic material
organisation.*

matForm abstracts the robot end-of-arm tool (EOAT) to operate as a single autonomous agent, moving in 3D graphical space, where it makes incremental decisions to define an AM tool path relative to its local perception of structural and geometric data gleaned from an analysis of a user-input 3D mesh model (e.g., a topoForm#1 column geometry). An initial set of horizontal contours first approximates the layered manufacturing of the geometry relative to an AM extrusion nozzle diameter. A robot tool path is then created by moving an EOAT

Additively Manufactured Architecture

6.17 TopoForm#2
Ceiling with truss-like
ribs in large ceiling
span locations.

extruder-agent (particle) through each contour layer, deviating locally to produce overhangs or other variations in trajectory and velocity that would result in material-extrusion gravitational effects (Figure 6.25). These variations add design interest but also prevent structural or manufacturing-caused collapse by ensuring sufficient material bonding in areas of high structural utilisation and limiting extreme trajectory deviations in areas with overhangs.

6.18 SituatedFabrications
Close-up of viscous
concrete material
deposition. Material
effects partly caused by
robot motion and partly
by viscosity of concrete.
Squiggles were the
result of the material's
viscosity and extrusion
rate while the overall
path was defined by the
robot's trajectory.

The extruder-agent follows the contours while adapting its motion in relation to locally perceivable geometric or structural data from the input mesh. It moves incrementally using vector steering behaviours, including steering towards the contour, aligning to the contour's tangent, and rotating away from its current velocity. These behaviours are weighted in relation to the agent's current state, and converted to a single acceleration vector that is added to the agent's velocity to determine its next position. The agent changes state in response to its previous states and time, such as:

if the agent was state 'A' for the last ten time-frames, change to state 'B'

or due to locally perceived structural or geometric data:

if the closest mesh face inclination angle is above -30°, change to state 'C'

Material-scale design features in *matForm* are created using sequences of vector moves determined by agent states and their corresponding steering behaviours. These moves result in waypoints with specified robot velocities (Figure 6.25c). To design these features, a bottom-up sequence of rules is developed, tested, and catalogued. Each is evaluated for its visual character and ability to register gravitational-material effects, with the most successful ones incorporated into the matForm algorithm. A matForm design is completed after the

▶
6.19 SituatedFabrications
a-b) MEX-AM concrete
panels (approx. 1.5m
in height) with interior
cavities.

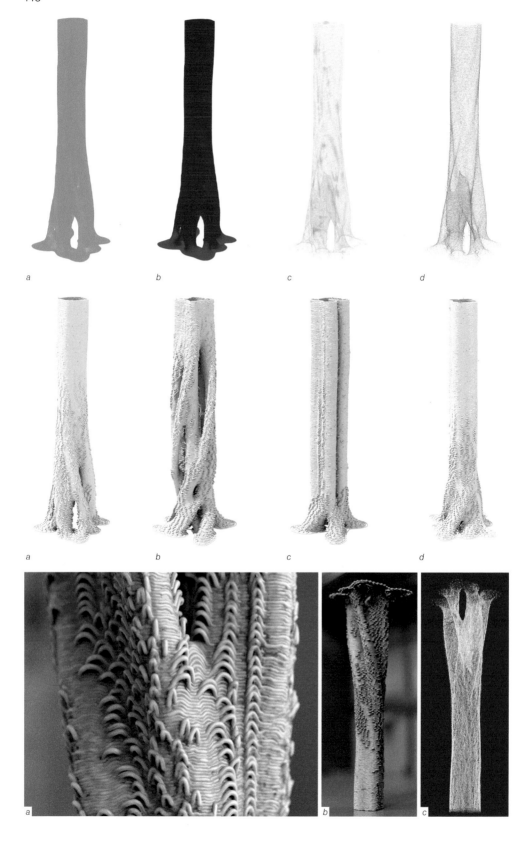

a   b   c   d

a   b   c   d

a   b   c

◄ **6.20 matForm**
Local geometrical and structural analysis data informs matForm agent behaviour. a) curvature, b) inclination, c) structural stress, d) matForm material placement program. Note that the column is shown inverted, the orientation in which it is manufactured.

◄ **6.21 matForm**
matSim results from matForm robot programs on four different column geometries (a-d) with an identical rule set. (a) is the same geometry as analysis images shown in Figure 6.20.

robot agent passes through all contours defining a new custom set of trajectory points and velocities. There is no randomness to the method. Repeated runs of the algorithm will produce identical results. Despite this, results are difficult to predict. While rules governing changes to the agent's state are deterministic, the agent's use of vector steering behaviours to adapt to local geometrical data is non-deterministic. The bottom-up generated robot tool path cannot be predicted prior to simulation, due to the robot agent responding to its current perception in addition to past experiences.

Applications of matForm and matSim
A selection of column-like outcomes from different *topoForm#1* evolutionary runs were used to demonstrate the capabilities of *matForm* and *matSim*. Each was inverted during simulated manufacturing to reduce overhangs that might cause collapse and implemented the same *matForm* rule sets. Using *matForm*, a simulated robot tool path was developed for each geometry. By mapping the changes in state of the robot's extrusion-agent with colour, the material deposition sequence can be visualised in a DNA-like manner (Figure 6.24). Rendered outcomes highlight *matForm's* ability to adapt design outcomes to the unique properties of each column geometry, resulting in diverse order based on a shared set of event-based, behavioural rules (Figure 6.21). Rendered images of *matSim* results also illustrate extensive variation in the character of material features and propagation effects (Figure 6.22a-b). The subtle and high-quality geometrical aspects seen in these images demonstrate *matSim's* usefulness as a design and visualisation aid. Structural analysis was also successfully performed on *matSim* outcomes[6], providing structural feedback on design solutions (Figure 6.22c). From the *topoForm#1* outcomes, two columns were fabricated in clay by MEX-AM, each using different variations of matForm rule sets (Figure 6.26). Figures 6.26a and 6.28 clearly exhibit design features closely related to the curvature and inclination of the column-like geometries. Additionally, variations in material behaviour stemming from the same set of rules are apparent. This is primarily due to the effects of inclination and curvature on gravitational settlement, as well as the incremental scaling of features along the height of the geometry.

◄ **6.22 matSim**
a) Close-up of a 3D rendered visualisation of a matSim simulated manufactured outcome with gravitational material effects and subtle propagation effects across layers, b) overall simulated outcome, c) high-resolution structural analysis accounts for variable inter-layer bonding.

Additively Manufactured Architecture

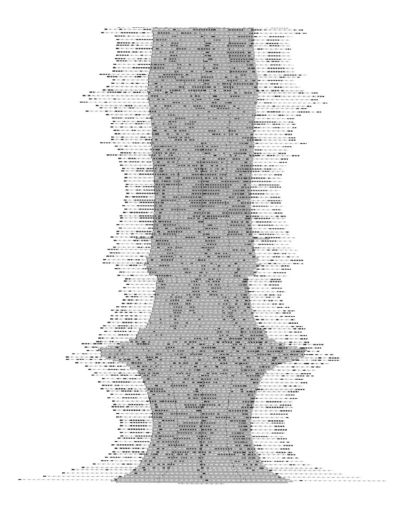

▲

6.24 matForm
An unwrapped diagram of matForm's behaviour-based rules for manufacturing a column in contour layers (shown as rows). Colours denote different robot velocities and degrees of deviation from the original geometry's contours.

Together, *matForm* and *matSim* offer a generative design approach to MEX-AM that is informed by structural, geometric and robot motion parameters. The method also provides visual and structural feedback on high-resolution material effects, extending design cognition into material and production concerns prior to manufacture. In *matForm*, robot motion was abstracted to a sequence of computed states that determined a tool path and velocity sequence. The method also creates an offline robot program that can support actual and simulated AM[7]. The method recasts an industrial robot as a semi-autonomous agent, making decisions based on prior actions and locally perceived data. This robot agent operated under event-based rules, generating material-production behaviours that tie the work of this chapter to previous chapters. While simpler methods can also produce AM material effects, the presented approach has the distinct advantage of being extendable to operate in several digital and material domains to encompass the diverse endeavours described in this book. This behaviour-based approach enables design agency to operate beyond a building's formal or organisational design, to determine material organisation and effects, or to engage in adaptive robotic fabrication or collective robotic construction activities described in subsequent chapters.

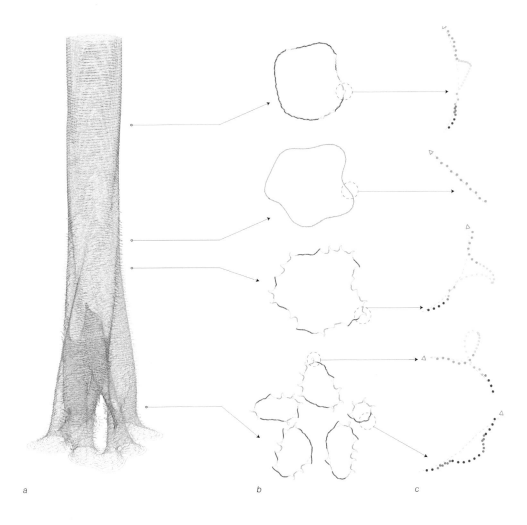

a   b   c

6.25 matForm
A column's matForm
robot trajectories
shown inverted for
manufacturing.
a) matForm contours,
b) individual matForm
contours, c) contours
consist of behavioural
sequences that deviate
from an original
reference contour
(shown dashed).

▶▶
6.26 matForm
Ceramic topoForm#1
column geometries
additively manufactured
with matForm-generated
material placement (a-b).

Incorporating Material Formation and Robot Manufacturing within Design

The previous two chapters presented methods for embedding design intent in 3D space using multi-agent algorithms. This chapter demonstrates how similar approaches can be extended to address additively manufactured architecture. It also shows how robot production and material placement can also be engaged to create manufacturing-scale material effects. Altogether, the work of this chapter explored an architecture that arises from generative approaches to topological and material formation. Agent behaviour operated in a computer algorithm to not only generate 3D geometric designs but also to create a program that could be executed on an industrial robot, governing its behaviour over time to affect a material production process. While this represents a computational approach being expanded to engage in physical world actions, most event-based decision-making operated inside of a virtual 3D graphical space. The chapters that follow extend this approach to more situated means of engagement with the world at large. First, through autonomous perception – bringing the observed world into 3D graphical space, and then onwards to the embodiment of computation within live robot behaviour in the physical world, before exploring the situated behaviour of several mobile robot systems undertaking collective robotic construction.

Additively Manufactured Architecture

a

Notes

1. *Manufactured in less than 48hrs, the project as a whole was completed in eight days. The company claims it can reduce the time and cost of building by as much as 70% relative to industry-established methods (Essop 2020).*

2. *Formative manufacturing includes precast concrete casting, injection moulding, rotomoulding, single-point forming (e.g, incremental metal forming), vacuum forming, rod and tube bending, press-braking, punch pressing and other methods that either deform or cast material.*

3. *Seek methods calculate a vector steering towards the average position of neighbours while avoid methods calculate an inverse vector. Align methods average the orientation of neighbours. Each method's vector result is scaled to weigh its relative influence over an agent's behaviour.*

4. *Rhino3D™ Grasshopper™ Galapagos evolutionary solver (Rutten 2013), was used in conjunction with a custom-developed multi-agent algorithm developed by the author.*

5. *Structural deflection and principal stress data is obtained from a structural analysis of a ceiling-slab mesh model with specified column and wall support locations. This data is then stored and locally accessed to create an initial distribution of agents, and used to influence agent motion behaviours.*

6. *Structural analysis was performed in Rhino3D™ Grasshopper™ using the Karmamba3D™ plugin.*

7. *matForm produces a sequence of waypoint planes (position and orientation) and associated robot velocity instructions. In this research, this data was provided to Grasshopper's Visose Robots plugin to produce an offline ABB robot program (Soler 2016), although, any one of several plugins for exporting industrial robot programs could be utilised.*

Project Credits

Center for Innovation and Precision Dentistry (2020): *Autonomous Manufacturing Lab, University of Pennsylvania (AML-Penn). Project Lead: Robert Stuart-Smith, Project Team: Patrick Danahy. Collaborators: Dr. Hyun (Michel) Koo, Penn Dental. Ellen Neises & Julie Donofrio, Penn Praxis.*

topoForm#1 (2020): *Autonomous Manufacturing Lab, University of Pennsylvania (AML-Penn). Research Lead: Robert Stuart-Smith, Researchers: Patrick Danahy, Natalia Revelo.*

topoForm#2 (2020): *Autonomous Manufacturing Lab, University of Pennsylvania. Research Lead: Robert Stuart-Smith, Researchers: Patrick Danahy, David Forero.*

viscera/L (2021): *Autonomous Manufacturing Lab, University of Pennsylvania. Research Lead: Robert Stuart-Smith, Researchers: Patrick Danahy, Renhu (Franklin) Wu.*

matSim (2020): *Autonomous Manufacturing Lab, University of Pennsylvania. Research Lead: Robert Stuart-Smith, Researcher: Sanjana Rao.*

matForm (2020): *Autonomous Manufacturing Lab, University of Pennsylvania. Research Lead: Robert Stuart-Smith.*

matForm manufactured prototypes (2023): *Autonomous Manufacturing Lab, University of Pennsylvania. Research Lead: Robert Stuart-Smith, Researcher: Renhu (Franklin) Wu.*

SituatedFabrications Fabrication Studio (2016). *Visiting Professor: Robert Stuart-Smith. Academic Chair: Marjan Colletti. Academic Staff: Johannes Ladinig, Georg Grasser, Pedja Gavrilovic. Students: Andreas Auer, Monique banks,*

Marcus Bernhard, Thomas Bortondello, Marc Differding, Christophe Fanck, Konstantin Jauck, Emanuel Kravanja, Jil Medinger, Sonia Molina-Gil, Pol Olk, Michael Schwaiger, Mario Shaaya, Melina Stefanova, David Stieler, Theresa Uitz, Matthias Vinatzer, Alexandra Zeinhofer.

Image Credits

Figures 6.1–6.16, 6.21–6.28: Robert Stuart-Smith, Autonomous Manufacturing Lab, University of Pennsylvania.

Figures 6.17–6.20: SituatedFabrications Fabrication Studio, Visiting Professor: Robert Stuart-Smith, University of Innsbruck

References

Aghaei Meibodi, Mania, Mathias Bernhard, Andrei Jipa, and Benjamin Dillenburger. 2017. "The Smart Takes from the Strong 3d Printing Stay-in-Place Formwork for Concrete Slab Construction." In Fabricate 2017: Rethinking Design and Construction, edited by Achim Menges, Bob Sheil, Ruairi Glynn, and Marilena Skavara. UCL Press.

Anton, Ana, P Bedarf, A Yoo, L Reiter, et al. 2020. "Concrete Choreography: Prefabrication of 3D Printed Columns." In Fabricate Making Resilient Architecture.

Apis Cor. 2017. "Apis Cor- We Print Buildings." Apis Cor.

Borsi, F, and P Portoghesi. 1996. Victor Horta. Belgium: Editions - J.-M Collet.

Boruslawski, Piotr. 2016. "Joris Laarman Lab 3D Printed Aluminum Gradient Chair." Designboom. March 29. https://www.designboom.com/technology/joris-laarman-lab-aluminum-gradient-chair-03-29-2016/.

Bromberger, Jörg, Julian Ilg, and Ana Maria Miranda. 2022. "The Mainstreaming of Additive Manufacturing."

Bromberger, Jörg, and Richard Kelly. 2017. "Additive Manufacturing: A Long-Term Game Changer for Manufacturers." McKinsey & Company. https://www.mckinsey.com/business-functions/operations/our-insights/additive-manufacturing-a-long-term-game-changer-for-manufacturers#.

Buswell, R. A., R. C. Soar, A. G. F. Gibb, and A. Thorpe. 2007. "Freeform Construction: Mega-Scale Rapid Manufacturing for Construction." Automation in Construction 16 (2). doi:10.1016/j.autcon.2006.05.002.

Caterpillar. 2022. "Caterpillar | Cat Reman." 2021. Accessed March 7. https://www.caterpillar.com/en/brands/cat-reman.html.

Corbusier, L. 2013. Towards a New Architecture. Dover Architecture. Dover Publications.

Dawkins, Richard. 1986. The Blind Watchmaker: Why the Evidence of Evolution Reveals a Universe Without Design. National Bestseller. Science. Norton.

Dreith, Ben. 2022. "Construction Commences on BIG and ICON's Community of 3D-Printed Homes in Texas." Dezeen. November 11. https://www.dezeen.com/2022/11/11/icon-big-wolf-ranch-3d-printed-homes-austin/.

D-Shape Enterprises. 2022. "D-Shape Enterprises L.L.C. – Global Mega-Scale Additive Manufacturing." Accessed March 12. https://dshape.wordpress.com/.

Eschenauer, Hans A., and Niels Olhoff. 2001. "Topology Optimization of Continuum Structures: A Review." Applied Mechanics Reviews. doi:10.1115/1.1388075.

b

Additively Manufactured Architecture

Essop, Anas. 2020. "SQ4D 3D Prints 1,900 Sq Ft Home in 48 Hours." January 13. https://3dprintingindustry.com/news/sq4d-3d-prints-1900-sq-ft-home-in-48-hours-167141/. ETHZ BRG and Zaha Hadid ZHACODE. 2021. "Striatus 3D Concrete Printed Masonry Bridge." November. https://www.striatusbridge.com/.

Finney, Alice. 2023. "Five Standout 3D-Printed Shoes Unveiled at Paris Fashion Week." Dezeen. February 3. https://www.dezeen.com/2023/02/03/3d-printed-shoes-paris-fashion-week/.

Ford, Simon, and Mélanie Despeisse. 2016. "Additive Manufacturing and Sustainability: An Exploratory Study of the Advantages and Challenges." Journal of Cleaner Production 137 (November). Elsevier: 1573–87. doi:10.1016/J.JCLEPRO.2016.04.150.

Forust. 2021. "Forust." https://www.forust.com/.

Frazer, John. 1995. An Evolutionary Architecture. An Evolutionary Architecture. London: Architectural Association.

Garcia, David, MacKenzie E. Jones, Yunhui Zhu, and Hang Z. Yu. 2018. "Mesoscale Design of Heterogeneous Material Systems in Multi-Material Additive Manufacturing." Journal of Materials Research 33 (1). doi:10.1557/jmr.2017.328.

Gargiani, R, A Bologna, and J Haydock. 2016. The Rhetoric of Pier Luigi Nervi: Concrete and Ferrocement Forms. Treatise on Concrete. CRC Press LLC.

Gibson, Ian, David Rosen, and Brent Stucker. 2015. Additive Manufacturing Technologies. Additive Manufacturing Technologies. doi:10.1007/978-1-4939-2113-3.

Gibson, Ian, David Rosen, Brent Stucker, and Mahyar Khorasani. 2021. Additive Manufacturing Technologies. Additive Manufacturing Technologies. Springer International Publishing. doi:10.1007/978-3-030-56127-7.

Guimard, H, M L C Leconte, Musée d'Orsay, P Thiébaut, (France), Musée des arts décoratifs et des tissus (Lyon), Réunion des musées nationaux (France), and Musée lyonnais des arts décoratifs. 1992. Guimard. Éditions de la Réunion des musées nationaux.

Hansmeyer, Michael, and Benjamin Dillenburger. 2013. "Digital Grotesque: Towards a Micro-Tectonic Architecture." SAJ - Serbian Architectural Journal 5 (2). doi:10.5937/saj1302194h.

Holland, John H. 1975. "Adaptation in Natural and Artificial Systems, University of Michigan Press." Ann Arbor, MI 1 (97).

Hull, Charles W. 1986. Apparatus for production of three-dimensional objects by stereolithography. U.S. Patent No 4,575,300. Washington DC: Patent and Trademark.

Im, Hyeonji Claire, Sulaiman AlOthman, and Jose Luis García del Castillo. 2018. "Responsive Spatial Print: Clay 3D Printing of Spatial Lattices Using Real-Time Model Recalibration." In Recalibration on Imprecision and Infidelity - Proceedings of the 38th Annual Conference of the Association for Computer Aided Design in Architecture, ACADIA 2018.

ISO/ASTM. 2021. "ISO/ASTM 52900:2021(E): Additive Manufacturing — General Principles — Fundamentals and Vocabulary." International Standard 5.

Jipa, Andrei, Mathias Bernhard, Benjamin Dillenburger, Mania Meibodi, and Mania Aghaei-Meibodi. 2016. "3D-Printed Stay-in-Place Formwork for Topologically Optimized Concrete Slabs." In 2016 TxA Emerging Design + Technology, edited by Kory Bieg, 96–107. internal-pdf://0960461026/Jipa-2017.pdf%0Ahttps://www.research-collection.ethz.ch:443/handle/20.500.11850/237082.

Jipa, Andrei, Mathias Bernhard, Mania Aghaei Meibodi, and Benjamin Dillenburger. 2016. "3D-Printed Stay-in-Place Formwork for Topologically Optimized Concrete Slabs." In TxA Emerging Design + Technology. doi:10.3929/ETHZ-B-000237082.

Khoshnevis, B., R. Russell, H. Kwon, and S. Bukkapatnam. 2001. "Crafting Large Prototypes." IEEE Robotics and Automation Magazine. doi:10.1109/100.956812.

Lengyel, E. 2012. Mathematics for 3D Game Programming and Computer Graphics, Third Edition. ITPro Collection. Delmar Cengage Learning.

Lipson, H, and M Kurman. 2013. Fabricated: The New World of 3D Printing. Wiley.

Louth, Henry, David Reeves, Shajay Bhooshan, and Patrik Schumacher. 2017. "A Prefabricated Dining Pavilion Using Structural Skeletons, Developable Offset Meshes and Kerf-Cut Bent Sheet Material." In Fabricate 2017: Rethinking Design and Construction, edited by Achim Menges, Bob Sheil, Ruairi Glynn, and Marilena Skavara.

Lowke, Dirk, Enrico Dini, Arnaud Perrot, Daniel Weger, Christoph Gehlen, and Benjamin Dillenburger. 2018. "Particle-Bed 3D Printing in Concrete Construction – Possibilities and Challenges." Cement and Concrete Research 112. doi:10.1016/j.cemconres.2018.05.018.

6.27 matForm
Detail photograph of matForm material effects.

Additively Manufactured Architecture

Martin, Holly. 2021. "Robotic 3D Printing System Builds Large, Lightweight Structures in Free-Space." *The Fabricator*. March 23. https://www.thefabricator.com/additivereport/article/additive/robotic-3d-printing-system-builds-large-lightweight-structures-in-free-space.

Meyer, Hannes. 1976. "Programs and Manifestoes on 20th-Century Architecture." In *The Journal of Aesthetics and Art Criticism*, edited by Monroe C. Beardsley, Ulrich Conrads, and Michael Bullock, 34:117–18. Cambridge: MIT Press. doi:10.2307/430599.

Mitchell, M. 1998. *An Introduction to Genetic Algorithms*. Bradford Books.

MX3D. 2020. "MX3D Bridge." *MX3D*.

Oxman, Neri, Steven Keating, and Elizabeth Tsai. 2012. "Functionally Graded Rapid Prototyping." In *Innovative Developments in Virtual and Physical Prototyping - Proceedings of the 5th International Conference on Advanced Research and Rapid Prototyping*. doi:10.1201/b11341-78.

Parkes, James. 2021. "First Tenants Move into 3D-Printed Home in Eindhoven." *Deezeen*. May 6. https://www.dezeen.com/2021/05/06/3d-printed-home-project-milestone-eindhoven/.

Pollio, Vitruvius, and M H Morgan. 1960. *Vitruvius : The Ten Books on Architecture*. New York: Dover Publications.

Rael, Ronald, and Virginia San Fratello. 2017. "Clay Bodies: Crafting the Future with 3D Printing." *Architectural Design*. doi:10.1002/ad.2243.

"Relativity Space." 2022. Accessed March 7. https://www.relativityspace.com/stargate.

Reynolds, C. W. 1999. "Steering Behaviors for Autonomous Characters." *Game Developers Conference*, 763–82. doi:10.1016/S0140-6736(07)61755-3.

Rutten, David. 2013. "Galapagos: On the Logic and Limitations of Generic Solvers." *Architectural Design* 83 (2): 132–35. doi:10.1002/ad.1568.

Sasaki, M, Toyoo Ito, and A Isozaki. 2007. *Morphogenesis of Flux Structure*. AA Publications.

Scott, Clare. 2016. "Dutch Students Create a Unique 3D Printed Metal Bicycle with Help from MX3D." *3DPrint.Com*. February 3. https://3dprint.com/118086/dutch-students-3d-printed-bike/.

Sevenson, Brittney. 2015. "Shanghai-Based WinSun 3D Prints 6-Story Apartment Building and an Incredible Home." *3D Design, 3D Printing*.

Soler, Vicente, Gilles Retsin, and Manuel Jimenez Garcia. 2017. "A Generalized Approach to Non-Layered Fused Filament Fabrication." In *Disciplines and Disruption - Proceedings Catalog of the 37th Annual Conference of the Association for Computer Aided Design in Architecture, ACADIA 2017*.

Soler, Vincente. 2016. "GitHub: Visose/Robots." https://github.com/visose/Robots.

Stuart-Smith, Robert. 2011. "Formation and Polyvalence: The Self-Organisation of Architectural Matter." In *Ambience'11 Proceedings*, edited by Annika Hellström, Hanna Landin, and Lars Hallnäs, 20–29. Boras: University of Borås.

———. 2018. "Approaching Natural Complexity - The Algorithmic Embodiment of Production." In *Meeting Nature Halfway*, edited by Marjan Colletti and Peter Massin, 260–69. Innsbruck University Press.

Stuart-Smith, Robert, Patrick Danahy, and Natalia Revelo la Rotta. 2020. "Topological and Material Formation." In *Proceedings of the 40th Annual Conference of the Association for Computer Aided Design in Architecture: Distributed Proximities, ACADIA 2020*. Vol. 1.

Takahara, Yui. 2014. "3D Printed Bike Frame by James Novak." *I.Materialise: 3D Printing Blog*. August 11. https://i.materialise.com/blog/en/redesigning-the-bike-frame-james-novaks-experiment-with-3d-printing/.

Tang, Ming, and Noah Shroyer. 2020. "Cast-in-Place Freeform Concrete with Big Area Additive Manufacturing Formwork." *International Journal of Architecture, Engineering and Construction* 10 (2). doi:10.7492/ijaec.2021.009.

Tibbits, S. 2017. *Active Matter*. MIT Press.

Wohlers, Terry, and Tim Gornet. 2015. "History of Additive Manufacturing - Wohlers Report 2016." *Wohlers Rep*. 2012.

Wölfflin, H, P Murray, and K Simon. 1966. *Renaissance and Baroque*. Cornell University Press.

XtreeE. 2022. "XtreeE | The Large-Scale 3d." https://Xtree.Com/En/. Accessed February 10. https://xtreee.com/en/.

6.28 matForm
Detail photograph of matForm material effects.

Additively Manufactured Architecture

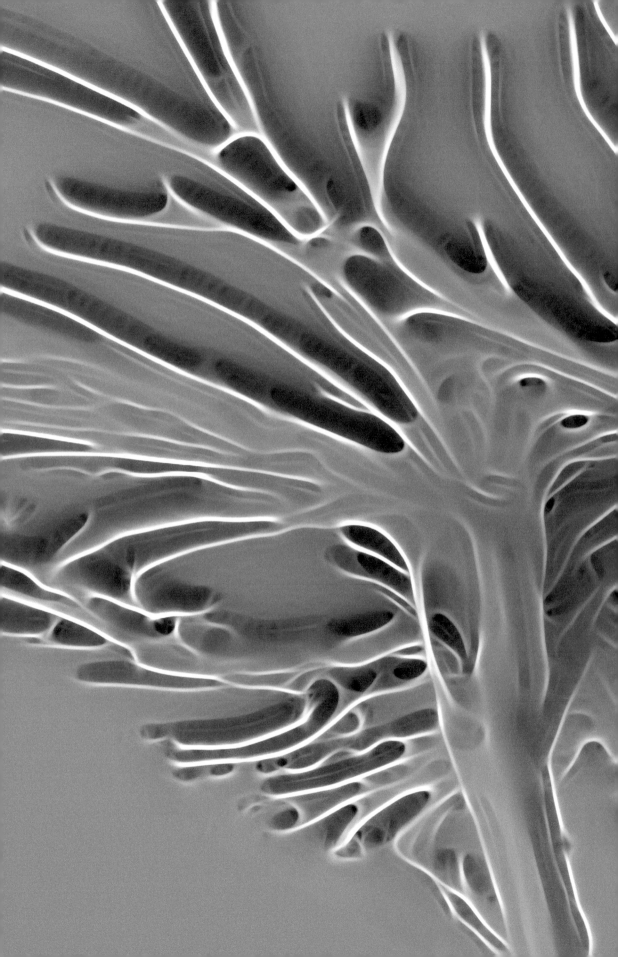

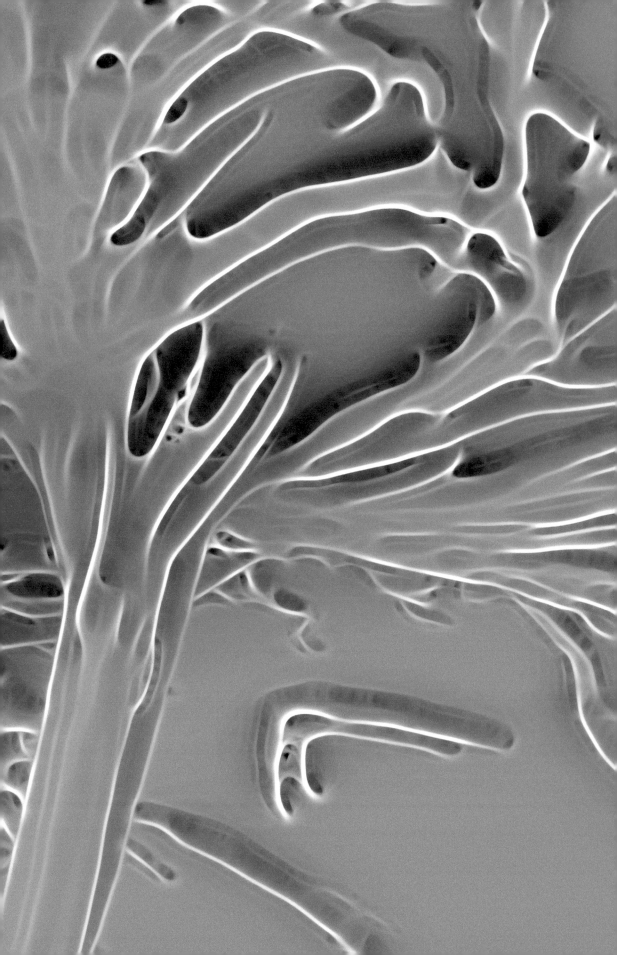

◀ ◀
7.1 Visual Character Analysis
This computer vision-generated image using Gabor Filtering is used to quantify one aspect of visual character: intricacy.

7 | Autonomous Perception
Leveraging Computer Vision and Machine Learning in Design

The act of design is heavily reliant on visual perception. In this chapter, autonomous modes of perception, analysis and generative design are explored that leverage computer vision and machine learning methods. The developed approaches consider user-defined and computer-generated visual character, while some of the work advocates for a temporal, situated architecture that challenges a disciplinary interest in difference to support greater specificity. The distinctiveness of an architectural work is also speculatively explored as a form of Posthuman architectural agency. The chapter commences by providing a disciplinary and technological context. Technical research is then presented on the development of various design methodologies and their corresponding outcomes. This is followed by a more discursive line of argument that explores the agency of architecture. A selective reader should feel free to jump between any of these sections at will, although there is additional meaning gained from the whole.

Architecture and Visual Representation

To engage with the world, we need some capacity to perceive it. As a spatial discipline, architecture has a long-standing dependence on visual perception and representation. In *The Projective Cast*, Robin Evans provides a historical account of significant architectural design developments that were influenced by advances in drawing and their translation into buildings. From Brunelleschi and Alberti's use of perspective (Evans 2000, 110–13), to orthographic projection methods in Andrea Palladio's San Giorgio Maggiore shallow depth façade, or the employment of stereotomic techniques of drawing and stone-cutting in Philibert DeLorme's Trompe D'Anet (Evans 2000, 117,186), Evans' examples illustrate how the representation of architecture has impacted design conception and construction. At present, substantial machine learning research is taking place in both academic and professional architectural settings into the use of Deep Neural Networks for the generation of visual media. The capacity for these methods to analyse and design various aspects of architecture is influencing new trajectories for architectural design and fabrication that will extend Robin Evan's discourse into currently unfathomable directions. While not discussed in this chapter, in the last twelve months alone, the rapid development of images from text prompt inputs in deep learning models such as *Midjourney*, *Dall-E* or *Stable Diffusion* (Islam 2022) has given rise to a vast dissemination of architectural imagery across social

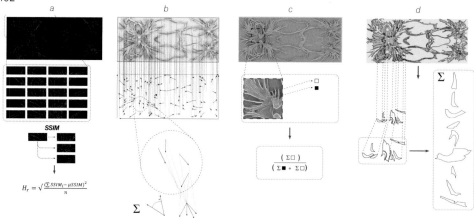

▲

7.2 Visual Character Analysis
Quantification methods:
a) heterogeneity, b)
continuity, c) intricacy, d)
number of recessed pockets.

media. The proliferation of this work is already impacting the education and practice of architectural design by challenging established workflows and notions of authorship and has been the subject matter of keynote discussions at leading conferences such as ACADIA 2022 (Akbarzadeh et al. 2023). Within this context, this chapter focuses on more established Deep Neural Networks (DNNs) applications for image content generation, image recognition, and reinforcement learning, and pursues custom approaches to their integration within 3D design work through extensive use of computer vision methods.

Deep Neural Networks (DNNs)

DNNs are a sub-field of Artificial Intelligence (AI) research that tasks a computer to learn from a predefined data set or from actual or simulated experiences. A DNN learns by adjusting the weighting of connections between data nodes distributed across a multi-layered data structure in order to produce information processing, pattern recognition or other problem-solving functions (Mitchell 2019, 21, Location 331). Deep Learning methods, such as Deep Reinforcement Learning (RL), Convolutional Neural Networks (CNNs) (Krizhevsky, Sutskever, and Hinton 2017), Generative Adversarial Networks (GANs) (Goodfellow et al. 2020) and others, enable weightings to be learnt or refined over time through several iterations of training. As a non-symbolic approach to learning, DNNs' weighted associations between nodes arises from billions of computational operations that render the logic of a trained DNN indecipherable to a human programmer. Even where a DNN operates on images, there is no visual intuition apparent of the logical data associations formed within a DNN (with perhaps the exception of a Neural Style Transfer CNN (Gatys, Ecker, and Bethge 2016)). Researchers are currently attempting to develop an alternative *"explainable AI"* that would facilitate human comprehension of a learned configuration (Mitchell 2019, 109, Location 1818). For now, DNNs are semi-autonomous and ambiguous, yet work incredibly well for some specifically trained tasks. Within the shrouded operations of several imaged-based applications of Deep Learning lies a vast sequence of computer vision image processing methods that are elevated to more prominence in this chapter, and directly leveraged to support various means of perception and action within design workflows, including the perception of architectural sites, and the strategic evaluation and incorporation of a designer's own intuitive, explicit and programmed contributions to design within Deep Learning workflows.

Computer Vision

In 1966, scientists Marvin Minsky and Seymour Papert initiated the Summer Vision Project, optimistically tasking a first-year undergraduate student with a summer research goal of enabling a computer to perceive and describe content from a television feed in the same way people do (Mitchell 2019, 69, Location 1117). Some fifty years later, the problem still seems considerably more complex than first assumed. To enable a computer to see like a human requires visual cognition, with the ability to recognise scenes, people, objects, and to separate or 'segment' these elements from an image solely comprised of an array of unrelated pixels, and to do so consistently across diverse lighting and shadow conditions (Mitchell 2019, 69, Location 1123). Computer vision researchers have developed numerous methods to support such efforts, many of which provide feature detection, pattern or object recognition. These require computer-programmed functions to operate on millions of pixels in every image. Due to several decades of advancement in graphics processors (GPUs), these tasks can now be parallelised and executed on miniaturised hardware that is easily integrated into cars, product designs or almost anything else (Davies 2017, xxv). Despite initially being associated with surveillance or security activities, this has enabled computer vision technologies to become increasingly ubiquitous across several aspects of our built environment, where the technology already supports the spatial awareness of robot vacuum cleaners and driverless cars to the facilitation of checkout-free AmazonGo™ corner stores[1] (Amazon 2023).

Although the research and development of computer vision itself sits primarily within the field of Computer Science, it is incorporated into the work of an expanding number of fields, including architecture. Examples include its use in architectural robotics assembly research (Tish, King, and Cote 2020) or in Nicholas Negroponte's MIT Architecture Machine Group's 1973 'Greet' project, where the software Greet monitored people passing through a door, performing edge detection procedures on a video feed to recognise and greet people by name (Negroponte 1975, 137). These examples demonstrate computer vision operating in the fabrication and operation of buildings. However, computer vision can also support architectural design, for both visual character analysis and generative design activities.

▼

7.3 Visual Character Analysis

Image of reflected ceiling plan showing different computer vision methods used for analysis: a) heterogeneity, b) intricacy, c) continuity, d) number of recessed pockets.

a   b   c   d

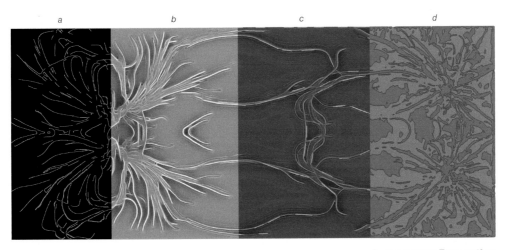

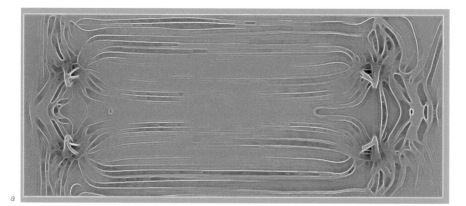

a

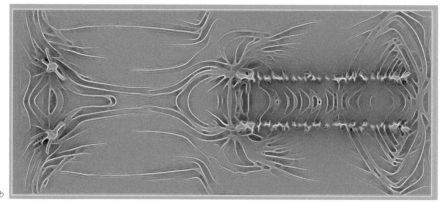

b

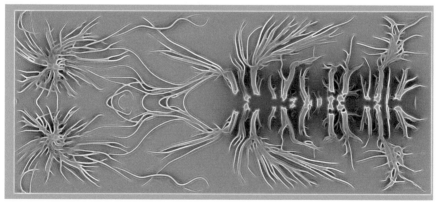

c

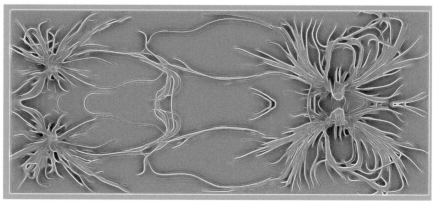

d

Visual Character Analysis (VCA) in Design

*7.4 Visual Character
Analysis*
Reflected Ceiling
Plans of four different
topoForm#2 designs
(a-d). Each image
uses Gabor filtering to
quantify intricacy.

In the previous chapter, a series of design approaches to additively manufactured architecture supported a reduction in material volume and the production of a distinctive aesthetic character[2]. Although designs were geometrically and structurally analysed, aesthetic considerations were not. Aesthetics were only visually assessed by designers during the coding and tuning of variables in algorithms prior to their use, requiring a designer to partner in the software development process to gain some aesthetic authorship. As aesthetics might also impact a design's material volume, there is a correlation between aesthetics and the carbon footprint and cost of a building, creating an environmental-ethical dimension that also needs consideration. Beyond broader aspects of aesthetics, the qualitative nature of a design's appearance is difficult to address within algorithmic design or optimisation methods that work with quantitative data, leaving a designer with limited means to reflect or quantify visual character's impact on material efficiency or other design constraints. To address this, an analytical approach to visual character was conceived to support the *Topoform#2* design methodology for additively manufactured architecture described in Chapter Six. In the Autonomous Manufacturing Lab (AML-Penn), a software called *Cyclops* was developed to provide viewport capture and image processing using the computer vision framework OpenCV (Bradski 2000) within the Rhino3D™ modelling environment (Stuart-Smith and Danahy 2022a). Computer vision methods were then utilised to translate aspects of a design's visual character into quantitative metrics that could be evaluated alongside structural or geometric analysis results.

Although the term 'visual character' could be used to describe a wide range of visible attributes, a narrow set of visual characteristics were identified that aesthetically related to the *TopoForm #2* generative design method and its tendency to produce multi-genus topological outcomes. These included heterogeneity, intricacy, continuity, and pockets (recesses). OpenCV functions were selected that could generate new images that corresponded to each characteristic, and methods were developed to quantify data from each OpenCV image outcome. The specific methodologies (Figures 7.2, 7.3) are described in detail in a paper published in CAADRIA'22 (Stuart-Smith and Danahy 2022a), and summarised below:

- *Intricacy:* greater amounts of geometrical definition. Gabor Filtering (Bradski 2000) is used to highlight edges in a geometry as white pixels, providing a quantifiable ratio of intricacy.

- *Heterogeneity:* greater degrees of difference between regions within an image. A Structural Similarity Index (SSIM) Image Difference comparison (Bradski 2000) of tiled regions within an image evaluates their similarity as a value of Heterogeneity.

- *Continuity:* degree of alignment between edges in areas of high curvature. OpenCV's Probabilistic Hough Lines Transform method is utilised to create a series of line segments that approximate continuous edges within an image (Bradski 2000). The angle between lines in proximity to each other is averaged to provide a metric of continuity.

- *Pockets:* degree to which a surface geometry is separated into regions of varying depth. An image mapping local 3D curvature is segmented into regions and categorised by colour using K Means Clustering (Bradski 2000). OpenCV contouring of clusters provides a count of clusters.

166

As discussed in Chapter Six, a matrix of *TopoForm #2* design outcomes was developed from several variations in spatial configuration and rule-set values (Figure 6.13). Outcomes were tested against visual character (from 3 camera views) together with structural and geometric analysis, to evaluate whether visual character criteria could be correlated with structural and material volumetric efficiency. As a proof of concept, the narrow range of visual character analysis methods' results did align with our aesthetic intentions. However, results also highlighted conflict between our preferences and material and structural performance. This suggests that the integration and weighting of visual character alongside structural and material considerations within a generative design process could facilitate a reduction in the material volume, cost and carbon footprint of architectural designs (Stuart-Smith and Danahy 2022a). In addition to analysis, *Cyclops* was also developed to support generative design activities within 3D modelling environments.

Incorporating Visual Character through Reinforcement Learning
As discussed in the previous chapter, recent advances in additive manufacturing (AM) allow architectural elements to be fabricated with increasingly complex geometrical designs and enhanced material (volumetric) and structural efficiencies. AM's ability to support one-off manufacturing on-demand also opens up more possibilities for user-customised design. Despite this, as Topoform#2 illustrates, developing designs that leverage AM's novel capabilities requires considerable knowledge and access to advanced software that is challenging for non-experts or people with disabilities to operate. Non-expert user participation in design has already been facilitated by the inclusion of machine learning (ML) within 3D design software in MIT's Architecture Machine Group in the 1970s (Negroponte 1972). Google Creative Lab™'s Autodraw™ also cognifies an inexperienced artist's scribbles to replace them with high-quality drawings (Motzenbecker and Phillips 2017). More

7.5 Visual Character Analysis
Two close-up views of a Topoform#2 design outcome (a-b), adjacent to corresponding intricacy VCA images: a perspective (c) and reflected ceiling plan (d).

recently, architectural research has incorporated ML to assist in the incorporation of performative aspects of user-created designs (Akizuki et al. 2020; Yousif and Bolojan 2021), or to assist in designing assemblies from a predefined set of elements (Hosmer et al. 2020). Established methods, however, do not allow a non-expert user to intuitively specify a 3D intention from multiple sourced or sketched 2D images to generate a multi-genus 3D topological model, nor do they attempt to preserve visual character intuited by the designer. Extensive research into computer vision and ML-based approaches to 3D reconstruction from several sourced images exists in the field of computer science (Soltani et al. 2017; Ham, Wesley, and Hendra 2019), yet these focus on voxel-based or mesh translation and do not incorporate architectural design-engineering workflows or attempt to embody visual character derived from a user's input images.

In the AML-Penn, we questioned whether we could extend the *topoForm#1* agent-based design methodology discussed in the last chapter that generates multi-genus topological column-like geometries, to develop its designs in response to user input images. The method – *topoFormVC* – attempts to safeguard a user's gestural or aesthetic intentions by evaluating the perceptual equivalence of 3D design outcomes relative to user-specified 2D image content within ML model training. Similar to *topoForm#1*, in *topoFormVC*, a series of agents are tasked to descend from a specified height within a defined Cartesian space where their motion trajectory is interpolated as a volumetric mesh. Agent trajectories attempt to match content from three user-input images, each associated with different 3D camera projections. Although these images do not necessarily corroborate a single design solution (their camera angles and content might not create a projected intersection), each agent's behaviour must do its best to approximate spatial configuration and visual character arising from all input images. This is extremely challenging. Without an ensured solution, the problem was an ideal candidate for Reinforcement Learning (RL), which is effective at resolving problems where solutions are not self-evident.

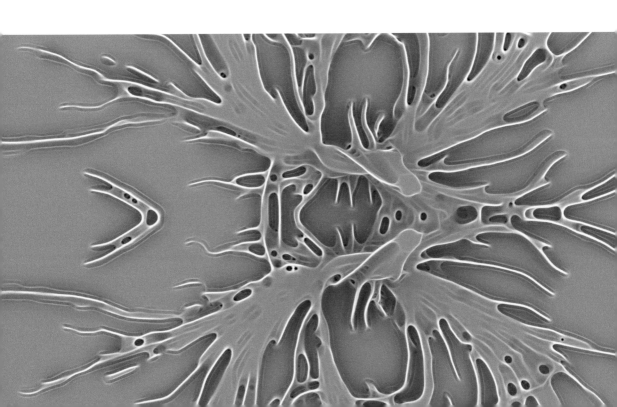

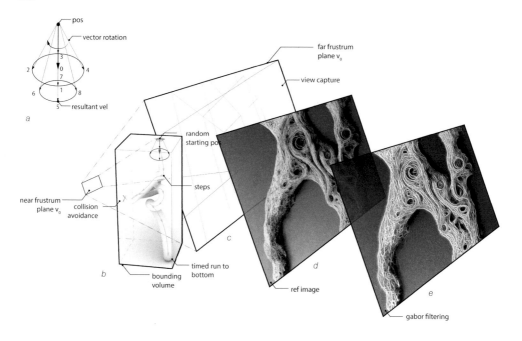

pos

vector rotation

3
2 — 0 — 4
7
6 — 1 — 8
5 — resultant vel

a

far frustrum plane v_0

view capture

random starting pos

steps

near frustrum plane v_0

collision avoidance

timed run to bottom

b

bounding volume

c

d

ref image

e

gabor filtering

7.6 topoFormVC
*Sequential Motion (SM):
a) SM agent motion, b)
SM environment set up
with camera views and
c) one camera's image
capture projected to its
far frustrum plane, d)
user-input reference
image and e) Gabor
filtered image used
in SSIM analysis of
generated outcomes.*

In *topoFormVC*, a behavioural strategy for each agent's motion trail is developed by a reinforcement learning process that modifies one agent's behaviour. A series of incremental agent decisions creates a sequential motion (SM) behaviour that is adjusted by a Reinforcement Learning (RL) process[3] (Chollet and others 2015) that trains agent steering behaviours over thousands of iterations (Figure 7.6). A reward function operates at every motion step within a simulation and penalises an agent for leaving a predefined volume, while rewarding it for continuing to the bottom of the volume or for increasing its topological genus (Stuart-Smith and Danahy 2022b). Most importantly, images of the simulation result are compared to user-input images from the same projection angles using OpenCV's Structural Similarity Indexing (SSIM) method (Wang et al. 2004) to ascertain their perceptive similarity (Figure 7.6).

While the RL process trains agent behaviour to better approximate user-input images, the agent is essentially flying blind until the SSIM training eventually steers towards improvements after thousands of simulations. To enhance the process, *Cyclops* computer vision analysis methods are also utilised to inform each agent motion step directly. A second agent-based locally adaptive motion (LAM) method remaps input image data into 3D space as a distributed field of vectors and voxels (Figure 7.7). Where projected data points from multiple images align, the data is summed to provide greater weighting. LAM agent behaviour seeks and aligns with this 3D data, enabling agent motion to directly follow image input data throughout each simulation. Using a combined SM and LAM agent behavioural approach within the RL model, agent motion is able to adapt directly to user-input image computer vision data in addition to being trained to improve perceptual similarity to user-input images. Although the research is a work-in-progress and does not currently produce column-like geometries, it demonstrates both organisation and aesthetic character are being approximated in generated outcomes (Figure 7.8), suggesting the method holds great promise (Stuart-Smith and Danahy 2022b).

The VCA and topoFormVC methods utilise computer vision to augment a generative design process within a 3D modelling environment. In parallel to these endeavours, *Cyclops* was also extended to support a similar design approach capable of engaging with more open, site-specific environments.

Cyclops: Integrating Machine Vision and Learning within Generative Design

Inspired by LeCun et al's *LeNet-5* Deep Neural Network (LeCun et al. 1998), machine-learning (ML) approaches such as Convolutional Neural Networks (CNNs) (Krizhevsky, Sutskever, and Hinton 2017) and Generative Adversarial Networks (GANs) (Goodfellow et al. 2020) have the capacity to rapidly process large amounts of visual data and unify design and computer vision tasks within one generative methodology. While 3D-GANs and other 3D ML algorithms are emerging (Wu et al. 2016), to date, architectural ML-based design research has primarily explored the use of image-based designs, or 3D designs derived from image-based ML that map 2D image pixel content to a fixed 3D mesh's colour and displacement maps, or interpolate several ML output images into a 3D voxel array (del Campo et al. 2021; Rehm 2019). Prior to recent Large Language Model (LLM) ML software such as Midjourney™ (Islam 2022) (not discussed in this chapter), several architectural works have employed StyleGAN (Klein 2020; Campo, Manninger, and Carlson 2019) or CycleGAN (Rehm 2019) to develop designs trained on architectural precedents or novel outcomes from data sets unrelated to architectural scenarios. The former essentially hybridises precedent works into compositions bearing some resemblance to David Griffin and Hans Kolhoff's *'City of Composite Presence'* drawing featured in Rowe and Koetter's book, *Collage City* (Rowe and Koetter 1983) yet with greater degrees of integration between disparate parts (that arguably go beyond collage). The latter conjures up exciting speculations on alternative architectures, where machinic perception produces an uncanniness architect Karel Klein describes as arising from *"unseen recollections of something external to ourselves"* (Klein 2022). The

▼

7.7 topoFormVC
Local Agent Motion (LAM): a) LAM environment set up with b) Canny edge and c) Hough Line analysis of user-input reference image data that is remapped into d) voxel and e) vector spatial data.

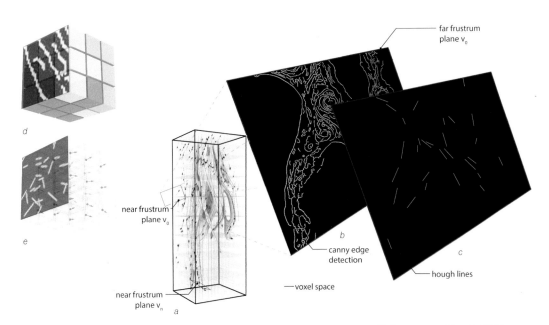

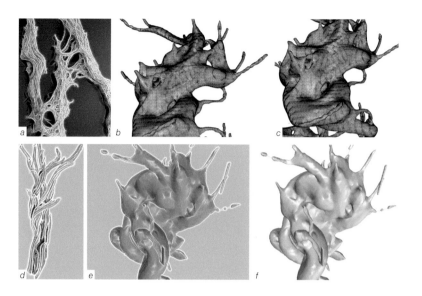

▲

7.8 topoFormVC
*Computer vision analysis
of user-input images
(a,d), adjacent to work-in-
progress results (b-c, e-f).*

success of these approaches has been contingent on a design outcome's obscurity or estrangement from architectural or input data rather than their engagement with specific 3D building sites. As designs are developed primarily through selective sourcing of 2D image data, approaches that also leverage a designer's existing algorithmic or 3D modelling workflows remain relatively under-explored. Substantial creative opportunity lies in incorporating the machinic perception of architectural sites and a designer's individual 3D design approaches within ML processes.

Geospatial mapping methods using LiDAR[4] or photogrammetry can create a textured 3D mesh or point cloud with sufficient information for a designer to develop 3D proposals that are spatially integrated into a building site without necessitating a physical site visit. Leveraging this, the AML-Penn developed software framework – *Cyclops* – enables a semi-autonomous design to be developed from 3D site scan data, using computer vision to extract site features, and multi-agent and machine learning methods to generate a design within the Rhino3D™ modelling environment (Figures 7.9, 7.10). *Cyclops* utilises the 3D model as the medium for all outputs and inputs throughout a multi-step design process. A designer is also able to algorithmically or explicitly model 3D geometries while engaging with computer vision and machine learning image-based processes within multiple camera views to perceive and further detail designs that are then incorporated into the 3D design model for use in subsequent steps.

OpenCV methods including "Canny Edge Detection" and "Hough Line Transform" are utilised to extract scanned site features including object contours and edges, with *Cyclops* translating their 2D raster results into lines in 3D Cartesian space. As each camera view contains noise (from shadows or over-exposed regions), images are semantically segmented prior to feature extraction to focus computer vision methods on desired features while ignoring irrelevant content. A Deep Neural Net developed by ENet with the *'Cityscapes'* trained data set (Paszke et al. 2016) is utilised to identify key architectural and urban features such as buildings, sidewalks, roads, lamp posts, windows, vegetation, people, sky, vehicles, etc. Extracted 3D point and line features are used as agent seeding locations and as vectors for agents to align to during a multi-agent design simulation (similar to topoForm #2 described in Chapter Six), allowing agent behaviour to spatially adapt to the 3D site environment. It is also possible for a designer to explicitly provide or modify such simulation inputs. A multi-agent simulation then produces a volumetric surface outcome that, due to its high-resolution and topological complexity, is difficult to modify using explicit 3D modelling methods. To incorporate more detailed design refinement, *Cyclops* employs two different machine learning processes; a Generative Adversarial Network (GAN) to add colour and texture to the design, and a Convolutional Neural Network (CNN) to add smaller-scale detail while preserving design character.

GANs consist of a generator model that is typically tasked with creating photorealistic image scenes and a discriminator model that tests how successful each outcome is. Together these develop novel image outputs over many computational cycles (Goodfellow et al. 2020). The creative abilities of GANs make them an excellent candidate for use in design; however, incorporating one within a broader building design methodology proves difficult as output images tend to diverge significantly from input data, resulting in outcomes that bear little resemblance to initial design inputs or a site context. Park et al's SPADE COCO GAN algorithm is perhaps an exception (Park et al. 2019). By operating only within semantically segmented regions of an image, it can on occasion preserve a degree of perspectival information, and other aspects of an original input image. In a series of specified 3D camera views, Cyclops semantically segments a multi-agent generated 3D mesh design model to ensure spatial, formal and aesthetic design character are preserved, and inputs the segmented images into the SPADE COCO GAN. Generated images provide additional visual and colour information and are brought back into the 3D mesh geometry assigning projected image pixel colours to the 3D design's mesh vertexes.

Ostagram™ (Morugin 2021; Ostagram 2021), a Neural Style CNN based on Gatys et al.'s *'Neural Algorithm of Artistic Style'* (Gatys, Ecker, and Bethge 2016) is then employed to produce new raster images of each 3D camera view to provide additional design detail. Gatys et al. demonstrate the method being utilised to apply the style and characteristics of artists such as Van Gogh and Kandinsky onto regular photographic scenes (Gatys, Ecker, and Bethge 2016). *Cyclops* explores a different approach, utilising two views of different zoom crops from the 3D design model (one image of the overall design, and another of a close-up detail) as inputs to produce a

▼

7.9 Cyclops
Cyclops is a custom software that employs camera views to both capture 3D data for 2D machine learning and image processing, and then can project results back into the 3D environment. a) multiple camera views within a 3D scanned environment, b) CV Hough Lines are used as input for a multi-agent generative design simulation c) outcome is CV analysed, segmented, and machine learning processes then generate final design data d) incorporated into the 3D model.

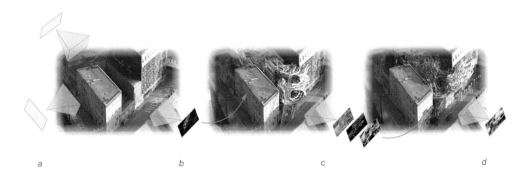

a   b   c   d

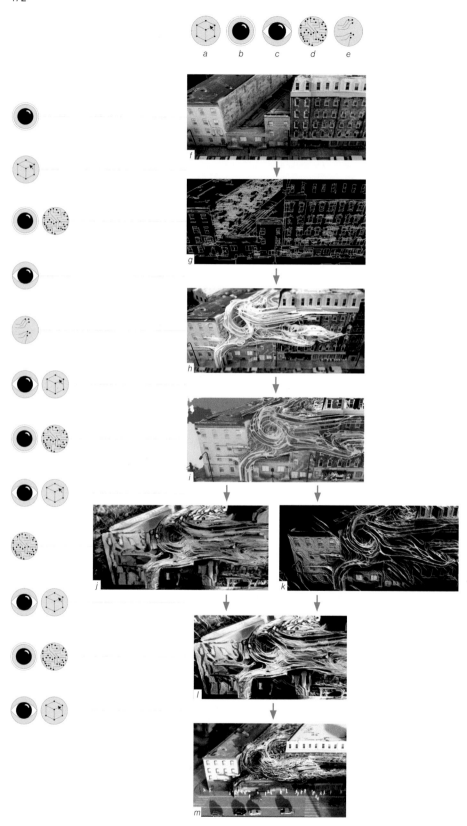

self-referential data set that generates increasingly complex and intricate variations of the original design while preserving its own content and style traits. Outcomes from the CNN are brought back into the 3D model, with image data translated to colour, opacity and displacement values on the geometry. Throughout a Cyclops design sequence, a designer still must make several decisions, guide each ML process, and can intervene if desired at any point to provide explicit 3D modelled input.

With these developments, *Cyclops* was able to semi-autonomously analyse a 3D site scan and extract site features for input into a multi-agent generative design method to enable designs to be site-specific. It could also employ a GAN and CNN to modify a design's colour, texture and opacity to provide additional finer-grain design detail. To evaluate its utility though, *Cyclops* needed to be tested on a design assignment.

Machinic InSites: Speculations on Semi-Autonomous Situated Design
Two iterations of a graduate architecture design studio at the University of Pennsylvania[5] explored architectural applications of the *Cyclops* method and provided a means to evaluate its design capabilities. The studio brief operated under the following speculative framework:

- *Urban Systems.* While the term "urban environment" conjures up images of public plazas, tree-lined streets, and sidewalks; these physical attributes of cities are now managed by less visible technological systems which operate with greater agency and frequency. Urban Operating Systems such as living-PlanIT™ (Living PlanIT AG 2021) manage city logistics in real-time, monitoring traffic, streetlight maintenance, or trash collection. The studio brief required a response to this present-day urbanism – a complex, chaotic system characterised by the interactions of a multitude of agencies, which are increasingly surveyed and managed by machine vision and learning technologies, of which we have limited knowledge or oversight.

- *Machine Vision/Learning.* The use of machine vision/learning technologies in cities raises concerns around individual privacy, racial biases, law enforcement practices, etc. Conversely, the same technologies have the capacity to be utilised for positive applications such as in medical diagnosis, environmental monitoring, autonomous vehicles, etc. The studio questioned how architects might utilise these technologies for constructive applications, to enable rapid forms of urban intervention, to augment design and manufacturing activities to deliver bespoke building solutions tailor-fitted to specific sites and manufactured in extremely short time-frames (such as by off-site additive manufacturing).

- *New Development Models.* While rapid building design and construction may not seem strategically advantageous for established means of development, it opens up alternative approaches to land use and ownership, such as the temporary occupation of under-utilised urban sites. A short life-cycle also encourages greater environmental accountability through cradle-to-cradle principles where short-term building supports materials being recycled and re-used on multiple sites.

- *Site Speculations.* Students imagined the use of additive manufacturing, temporary building and engagement with semi-

7.10 Cyclops
Activity Key:
a) explicit 3D modelling,
b) computer vision, c)
human visual assessment,
d) machine learning, e)
multi-agent simulation.

Cyclops design workflow
for Machinic InSites #1:
f) 3D site scan, g)
cv hough-lines, h)
multi-agent design, i)
semantically segmented
design, j) SPADE-COCO
GAN colour mapping, k)
Ostagram displacement
and opacity mapping,
l) final design outcome.

a

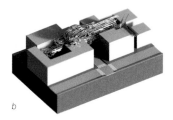

b

c

a

b

autonomous design processes as the premise under which to develop speculative design proposals for a rather vast and under-utilised site in Philadelphia's inner-city Callowhill neighbourhood, where a disused rail viaduct will soon be developed to form a section of Philadelphia's Rail Park (Friends of the Rail Park 2020).

Machinic InSites #1: Project Responses
Students designed infill project proposals, tailor-fitted to part of the site, and speculated on community-orientated urban entrepreneurialism, embracing the site's imminent transformation in different ways, each conceived to operate within a temporary leasing period of 5-20 years. One project addressed the disruptive urban transformation and gentrification wrought by development by proposing a museum that would operate as a neighbourhood time capsule. The museum would persist throughout various transformations of the site over time, through periodic partial demolition and extension, weaving between and around the site's developmental changes. The design embodies no centre or hierarchical organisation of space or structure. A decentralised, collective of spaces and material tightly adapts to the various permutations of site it lies within (Figures 7.10, 7.11, 7.12a).

A second project brought public programming to the site and addressed the recent closure of several employee rest spaces operated by the ride-sharing company, Uber™, proposing a public-private partnership. The proposal speculated that Uber™ might obtain lease-free use of city land for its rest spaces if it covered the cost of building and maintaining a local community art gallery space and a terraced plaza connection to the future viaduct section of the Rail Park. The project was envisaged to be built in phases coinciding with the Rail Park development schedule (Figure 7.13). A third scheme was grafted onto an existing school building, providing spaces to support the botanical management of the Rail Park while creating a spatial link between the school and the future park. The addition also speculatively operates as a plant substrate, decaying over time as plant growth expands from an indoor atrium garden into the Rail Park and beyond (Figure 7.15).

The studio's projects utilised Cyclops to engage with 3D site scan data through semi-autonomous methods for site analysis and generative design (Figures 7.10, 7.16). The methodology offered an opportunity to explore alternative forms of architectural agency, questioning how architecture might operate tightly within a site and help positively participate in temporal, spatial, and programmatic initiatives within the local community. The effectiveness of the

7.11 *Machinic InSites #1*
Neighbourhood time capsule building massing: a-c) adapts to neighbourhood development.

generative design approach is most apparent in the granular specificity of each design when viewed without the site, where a fine-grain fuzzy boundary to each design demonstrates how each is intimately and spatially connected to its 3D site (Figure 7.14).

Machinic InSites #2

Machinic InSites #1 projects successfully incorporated computer vision and machine learning methods within a generative design workflow and offered compelling speculations on architectural application scenarios. While the projects demonstrated greater visual complexity than work presented in earlier chapters, they were primarily developed using similar multi-agent design methods as were described in Chapters Four to Six. GAN and CNN methods operated solely within design-refinement activities to propagate a design character intrinsic to the multi-agent 3D mesh outcomes. However, these machine learning methods have the capacity to generate novel images that exhibit qualities or characteristics that arise from their non-human means of perception. A second iteration of the studio, *Machinic InSites #2*[6] sought to produce proposals that embodied more unique characteristics. While projects were to be developed for a specific site, they were to remain more object-like in their relationship with the site, with some sense of autonomy as temporal architecture.

In this second iteration, the studio jettisoned the multi-agent design and SPADE COCO GAN methods. Students commenced with the same 3D site scan StyleGAN data and *Cyclops* computer vision analysis methods but were taught to utilise Kerras et al's StyleGAN (Karras, Laine, and Aila 2018) to generate initial design outcomes. StyleGAN operated similarly to the GAN description provided earlier, developing novel images within a latent space generated using a pre-trained data set of images. The success of the method is largely determined by the quality and type of image content the GAN is trained with. To develop 2D image results that could be used to conceptualise 3D design proposals, students were tasked with 3D modelling conceptual 3D massings within their 3D site model in parallel to compiling a unique data set of images of subject matter related to their individual response to the project brief. Several hundred to thousands of images of these 3D design massings were also captured from a single camera view (or subtle variations of

7.12 *Machinic InSites #1*
3D partial building chunks. a) Neighbourhood Time Capsule project, b) Uber Plaza project.

7.13 *Machinic InSites #1*
Top roof-plan views of staged development for Uber Plaza proposal (a-c) for Uber rest space, community art gallery and public plaza.

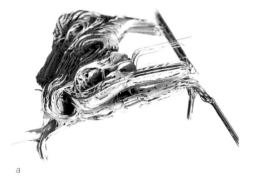

a

b

c

d

it) and added to a compiled image data set consisting of up to ten thousand images. All images (both modelled and scraped from the web) were reformatted to the same size with a white background. With an aspiration for the StyleGAN to train towards the production of objects rather than full-bleed images, only images of complete objects were submitted to the database.

The StyleGAN was then trained to generate novel images of potential 3D massings (Figure 7.17) that bore some relation to site via the explicitly modelled massing images within the data set, yet also appeared novel and not overly aesthetically constrained to the explicit models due to the other subject matter included in the training image set. Developing a successful proposal required several iterations, with changes to both 3D-modelled content and the image database. Results deemed successful in relating to both site and design intent were then explicitly modelled to match the GAN-generated images in their corresponding 3D model camera views. Design refinement was undertaken in a similar manner to *Machinic InSites #1,* using the same Neural Style CNN (Ostagram™) to develop a custom design articulation for each project.

As expected, *Machinic InSites #2* produced substantially more varied design outcomes compared to *Machinic InSites #1*. As with the previous design studio, students speculated on an entrepreneurial approach to improving the neighbourhood through temporal architectural interventions. One speculative proposal addressed a lack of mental health facilities in the neighbourhood by providing a new centre for Art Therapy. Art was envisaged to be produced in response to the building exterior where a synthetic nature alters the character of the existing neighbourhood (Figure 7.18). A less pronounced interior supports this, providing classes, an exhibition space, shop and various amenities (Figure 7.19). Another project proposed a temporal extension to a school building that incorporated exhibition and work spaces dedicated to the promotion of recycling. The project was to be manufactured from waste, upcycled into additively manufactured structural parts, clad with a robotically woven fabric envelope (Figure 7.20).

In each of these proposals, an architectural idea was both locally situated and aesthetically pointed. Designs were developed through a

◀

7.14 Machinic InSites #1
*(a-d) Aerial view of
project outcomes
shown without 3D site
context. The filagree
extensions to each
project demonstrates
site specificity
informed by Cyclop's
situated autonomous
perception methods.*

messy back-and-forth process where the students maintained a sense of authorship and agency in their work not only through explicit 3D modelling but also through the curation of a project-related image data set that indirectly steered the StyleGAN generative process. In these projects, aesthetics was not so much a question of style, as a question of content and haecceities (thisness). While all projects conformed to the same design methodology, no particular stylistic expression had to be maintained across projects, rather a unique character was sought for each project that was a sum of its architectural and subject-related influences.

Architectural Haecceities

In *Machinic inSites #1*, a multi-agent behavourial design approach adapted to different site conditions to produce diverse formal, spatial and aesthetic conditions, with design outcomes remaining characteristically consistent. In contrast, in *Machinic inSites #2* each project was distinctive and embodied a set of qualities and characteristics that emerged from unique ML-trained data sets. This distinctiveness, however, seemed to offer less spatial or formal dialogue with the building site. Version #1 was hyper-tailored to site, while #2 partially pulls back from this level of engagement in order to develop distinctive object-like outcomes. A hybridisation of these two approaches would offer more than either does individually. On reflection, altogether, the work discussed in this chapter raises questions about the possibilities and ideology that might be entailed in an architecture orchestrated through semi-autonomous machinic processes.

Equally to any other field, the discipline of architecture has developed in lockstep with technological developments. Following the second industrial revolution (IR2), architecture's embodiment of Fordist production principles was instrumental to its tectonic, aesthetic and political expression. In particular, the adoption of repetitive modular building elements, associated with aspirations to raise the bar on the build quality, hygiene and affordability of architecture, can also be criticised for reducing artisanship and aesthetic diversity in the built environment. With the third industrial revolution (IR3) introducing novel information technology infrastructure, buildings could be mass-customised. Several seminal works by Frank Gehry and others vary the geometry of building systems and parts (Brayer 2015). While supporting

▶

7.15 Machinic InSites #1
*Botanical management
spaces connect a rail park
to a school building. The
connection is envisaged
to operate as a substrate
for plant growth.*

Autonomous Perception

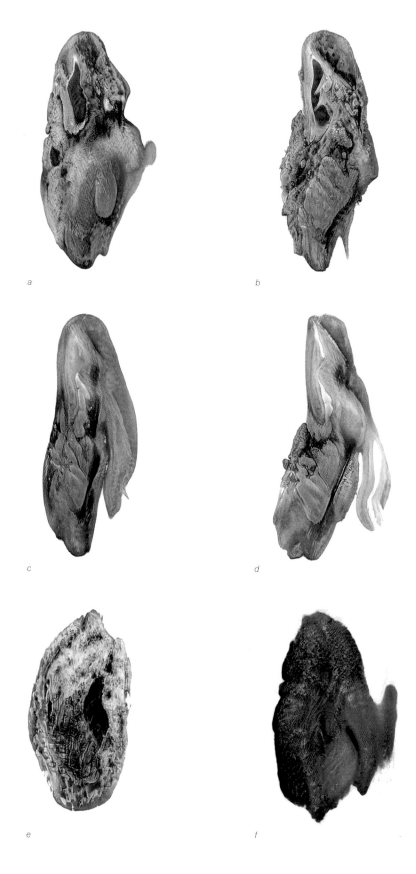

a

b

c

d

e

f

◀◀

7.16 Machinic InSites #1
Clockwise from top-left:
Multi-agent site-adapted
design, semantically
segmented view of multi-
agent design, SPADE-
COCO GAN generated
image, Ostagram
generated image.

more diverse architectural outcomes, IR3 designs are essentially variable forms of IR2 architecture that typically require a large budget and geometry rationalisation activities to support value engineering. What might the Fourth Industrial Revolution (IR4, discussed in Chapter Two) give rise to? Autonomous perception, design and manufacturing support more diverse and situated means of design engagement. Greater specificity in design is possible in relation to site, user input/ requirements, or a multitude of other considerations that might provide architecture with more agency and new aesthetic possibilities. This specificity needs a new discourse that moves beyond difference.

Difference through repetition, contradiction, smooth transition, or syntactical defamiliarisation has been explored within architectural projects and discourse for decades in various formats. In the last half-century alone it appears in Postmodernism, Deconstructivism, and the 1990s Digital design movements, and was inspired by philosophers including Jacques Derrida and Gilles Deleuze (Deleuze and Guattari 1988). In *Complexity and Contradiction*, Robert Venturi argued for juxtaposition and contradiction in response to the complexities of architecture, site and time (Venturi and Scully 1977). Peter Eisenman's early work challenged the syntactical structure of architecture (Eisenman 2018), while through the use of animation software and other computational methods, Greg Lynn and others sought complexity through smoothness and curvature (Lynn, Arts, and Kelly 1999). In Chapter Four, heterogeneous difference was also discussed in the context of complexity theory and emergent systems. Although more diverse than works from the International Modernist style that preceded them, projects from the above-mentioned periods are also more reductive than architecture developed within Baroque, Art Nouveau or even Classical or Renaissance periods. Despite their seminal contributions to the architectural canon and discourse, they are constrained by both their means of design and production, with most artisanal practices not extended into today's industrial means of construction.

◀

7.17 Machinic InSites #2
Top view of styleGAN-
generated designs, each
embodying different
haecceities (a-f).

As several chapters in this book advocate, in the emerging Fourth Industrial Revolution, artisanal-like bespoke work is now theoretically possible to achieve at industrial scales of production (with the next chapter exploring this to some degree). In *Machinic InSites #1 & #2* a form of situated bespoke architecture was explored, envisaged to exist on distinctively individualised terms, developed by human designers who leveraged semi-autonomous, machinic forms of perception, learning and manufacturing (at a conceptual level). The fact that this architecture was imagined to be recyclable and temporal further heightened its ability to be distinctive and one-off. These works set out to embody *haecceities* – with each project operating as a unique object exhibiting *'thisness'* (Ingram 2018).

The thirteenth-century CE philosopher-theologian-priest John Duns Scotus proffered the first known use of the term *'haecceity'* to describe a non-qualitative property that defines something as unique and indivisible – the *"thisness"* as opposed to the *"whatness"* of an object or substance (derived from the Latin *"haec"*, meaning *"this"*) (Cross 2022). In *The Rise of Realism*, philosophers Manuel DeLanda and Graham Harman agree that the term *'haecceities'* describes the unique identifying features of a specific cat, as opposed to things that could be shared with other cats (such as its breed, species or name) (DeLanda and Harman 2018, 8, location 271; Harman 2018). Although both Realists, DeLanda's philosophy is Neo-Materialist and heavily influenced by the work of Gilles Deleuze, while Harman is a founder of *Object Orientated Ontology (OOO)* that

Autonomous Perception

▲

7.18 Machinic InSites #2
Perspective view of Art
Therapy design proposal.

considers Materialism as a form of reductionism (DeLanda and Harman, 2018, 8, location 271; Harman, 2018). Despite their different philosophical dispositions, Harman and DeLanda both consider haecceities as necessary to ensure objects maintain a singular identity and cannot be reduced to a set of external relations or properties. However, for DeLanda objects are also *"unique historical individuals"* that embody *"processes of genesis and maintenance"* (DeLanda and Harman 2018, 52, location 1039).

Taken together, Harman and DeLanda's arguments can be appropriated to consider a distinctive work of architecture as an object (with exterior and interior conditions). If this work's haecceities are indivisible, its constituent parts are most likely unique. Architecture's part-to-whole relations are likely to remain of critical importance, but perhaps offer kinship to baroque and art nouveau work where these relationships embody some degree of ambiguity. Building on DeLanda's position further, an architectural work's haecceities can also be derived from its process of genesis and maintenance, which arguably consists not only in its design conception, but in the material and energy flows that brought forth its manufacture, and its ongoing physical interactions. Under these dynamic terms, what is the agency of such an architecture?

A Posthuman Architectural Agency
The rapidity of machinic processes such as computer vision or machine learning coupled with robotic means of production suggests that some form of architectural agency can be explored within alternative forms of temporality, where architecture might be more attuned to spatiotemporal concerns, offering a highly specific form of architectural space in contrast to Mies van der Rohe's advocation for flexible, universal space (Schulze and Archive 1985, 261). If Mies' clear-span architectural works metaphorically provided a stage, this temporal architecture creates a stage set, facilitating specific activities and provoking creative appropriation in lieu of neutrality. This impermanence liberates architecture from a need to provide functional flexibility, affording it new agencies. The emphasis here is not on the autonomy of the design process, but the autonomy of architecture itself – its material agency (see Chapter Three).

In *Vibrant Matter: A Political Ecology of Things*, political theorist Jane Bennett (who participated in the final jury review for *Machinic In-Sites #1*) brings to the reader's attention *"the Force of Things"* where she highlights the ability of human-made things to *"exceed their status as objects to manifest traces of independence or aliveness"* (Bennett 2010, Location 218). Speculative Realist philosopher Ian Bogost points out that human perception is only one of many ways objects relate to each other (Bogost 2012, 9). In citing Timothy Morton's *Ecology Without Nature*, which argues we should no longer perceive nature as a unified whole separate from society or human activity (Morton 2009, 181–97), Levi Bryant also suggests that for philosophical purposes, there is no world/container/whole. Just objects (including humans) that relate to each other by different degrees within different collectives (Bryant 2011, 270–72). Recasting architecture in this light (as the author argued in a panel discussion together with Bogost, Bryant and Morton in 2017[7]), requires additional ethical and creative considerations where the design, fabrication and final artefact of architecture must also engage with non-anthropocentric considerations, embracing Posthumanism. As Cary Wolfe conveys,

"Posthumanism means not the triumphal surpassing or unmasking of something, but an increase in the vigilance, responsibility, and humility that accompany a living world so newly, and differently, inhabited" (Wolfe 2010, 47).

In this context, while architecture could be more distinctive, the way its specificity is conceived, fabricated, instantiated, and appropriated could proffer greater levels of consideration for, and participation with, the world at large.

Towards Situated Design Responses

In this chapter, computer vision and machine learning methods were employed to support the semi-autonomous perception, analysis and design of buildings and building elements in relation to material and structural performance methods presented in the previous chapters, and in response to 3D scanned building sites and user-provided image content. Autonomous visual cognition of a building site also enabled design outcomes to be tailored to a site's physical environment. While these

▼
7.19 Machinic InSites #2
Perspective of 3D partial chunk of Art Therapy design proposal.

7.20 Machinic InSites #2
"Upcycle" project aerial perspective made from upcycled additively manufactured parts.

designs were conceived within a virtual encapsulation of the physical realm, their impact would only be realisable through physical construction, where agency would operate in each situated design artefact. The next chapter extends this situated condition into the physical world of robot manufacturing where several additional factors contribute to, and impact, design possibilities. Beyond this chapter's consideration for autonomously perceiving the physical environment within design conception, the next chapter explores adaptive forms of manufacturing, where design involves physical engagement. Embodied within robot platforms, design methods leverage robot sensing, tooling and actuation capabilities to creatively contribute to highly specific, bespoke design outcomes.

Notes

1. In an AmazonGo store a shopper's card details are obtained on entry to the store. A shopper does not go through a checkout to scan items, computer vision tracks what items they pick up and they are charged simply by walking out of the store with the items (Amazon, n.d.).

2. Chapter Six documented multi-agent design approaches to structurally optimised multi-genus topological designs of columns, walls, ceiling/slabs suited to material-extrusion-based additive manufacturing (MEX-AM). For more fundamental descriptions of multi-agent design processes refer to Chapter Four.

3. The research utilised Keras™'s RL API for TensorFlow 2 (Chollet & others, 2015), with a Functional Model comprised of several ReLU dense layers.

4. LiDAR is an acronym for Light Detection and Ranging

5. University of Pennsylvania M.Arch 701 Design Course, Studio Stuart-Smith. Instructor: Robert Stuart-Smith, Teaching Assistant: Patrick Danahy

6. University of Pennsylvania M.Arch 701 Design Course, Studio Stuart-Smith. Instructor: Robert Stuart-Smith, Teaching Assistant: David Forero

7. Deep Vista Conference, Texas A&M University. Chaired by Gabriel Esquivel, 2018.

Project Credits

Visual Character Analysis (2022): *Autonomous Manufacturing Lab, University of Pennsylvania (AML-Penn). Research Lead: Robert Stuart-Smith, Researcher: Patrick Danahy.*

topoFormVC (2022): *Autonomous Manufacturing Lab, University of Pennsylvania (AML-Penn). Research Lead: Robert Stuart-Smith, Researcher: Patrick Danahy.*

Cyclops (2020): *Autonomous Manufacturing Lab, University of Pennsylvania (AML-Penn). Research Lead: Robert Stuart-Smith, Researcher: Patrick Danahy.*

Machinic In-Sites #1 (2020): *University of Pennsylvania M.Arch 701 Studio Stuart-Smith. Students: Ximing Du, Chuanqi Gao , Sijie Gao, Huajie Ma, Heyi Song, Wenli Sui, Saina Xiang, Jing Yuan, Mingyang Yuan, Haochun Zeng, Yuhao Zhang, Zhe Zhong. Instructor: Robert Stuart-Smith, Teaching Assistant: Patrick Danahy.*

Machinic In-Sites #2 (2021): *University of Pennsylvania M.Arch 701 Studio Stuart-Smith. Students: Dongqi Chen, Hadi El Kebbi, Maria Sofia Garcia Perez, Michael "Benjamin" Hergert, Yaofang Hu, Liam Lasting, Anna Lim, Yulun Liu, Umar Mahmood, Danny Ortega. Instructor: Robert Stuart-Smith, Teaching Assistant: David Forero.*

Image Credits

Figures 7.1–7.8: *Robert Stuart-Smith, Autonomous Manufacturing Lab, University of Pennsylvania (AML-Penn).*

Figures 7.9–7.10: *Robert Stuart-Smith, Autonomous Manufacturing Lab, University of Pennsylvania (AML-Penn). Images used in the diagram from University of Pennsylvania M.Arch 701 Studio Stuart-Smith. Students: Sijie Gao, Zhe Zhong. Instructor: Robert Stuart-Smith, Teaching Assistant: Patrick Danahy*

Figures 7.11, 7.12a, 7.16: *Machinic In-Sites #1: University of Pennsylvania M.Arch 701 Studio Stuart-Smith. Students: Sijie Gao, Zhe Zhong. Instructor: Robert Stuart-Smith, Teaching Assistant: Patrick Danahy*

Figures 7.12b, 7.13: *Machinic In-Sites #1: University of Pennsylvania M.Arch 701 Studio Stuart-Smith. Students: Huajie Ma, Jing Yuan. Instructor: Robert Stuart-Smith, Teaching Assistant: Patrick Danahy*

Figure 7.15: *Machinic In-Sites #1: University of Pennsylvania M.Arch 701 Studio Stuart-Smith. Students: Haochun Zeng, Heyi Song. Instructor: Robert Stuart-Smith, Teaching Assistant: Patrick Danahy*

Figure 7.14: *Machinic In-Sites #1: University of Pennsylvania M.Arch 701 Studio Stuart-Smith. Instructor: Robert Stuart-Smith, Teaching Assistant: Patrick Danahy. Students: a) Sijie Gao, Zhe Zhong, b) Saina Xiang, Yuhao Zhang, c) Huajie Ma, Jing Yuan, d) Haochun Zeng, Heyi Song*

Figure 7.17: *Machinic In-Sites #2: University of Pennsylvania M.Arch 701 Studio Stuart-Smith. Students: Umar Mahmood, Hadi El Kebbi. Instructor: Robert Stuart-Smith, Teaching Assistant: David Forero*

Figures 7.18-19: *Machinic In-Sites #2: University of Pennsylvania M.Arch 701 Studio Stuart-Smith. Students: Yulun Liu, Dongqi Chen. Instructor: Robert Stuart-Smith, Teaching Assistant: David Forero*

Figure 7.20: *Machinic In-Sites #2: University of Pennsylvania M.Arch 701 Studio Stuart-Smith. Students: Anna Lim, Danny Ortega. Instructor: Robert Stuart-Smith, Teaching Assistant: David Forero*

References

Akbarzadeh, Masoud, Chigozie Nri, Joel Simon, and Kyle Steinfeld. 2023. "Origins and Destinations Beyond Midjourney: Keynote Presentation and Panel Discussion." In Proceedings of the 42nd Annual Conference of the Association of Computer Aided Design in Architecture (ACADIA): Hybrids and Haecceities, edited by Masoud Akbarzadeh, Dorit Aviv, Hina Jamelle, and Robert Stuart-Smith, 768–77. Philadelphia,: IngramSpark.

Akizuki, Yuta, Mathias Bernhard, Reza Kakooee, Marirena Kladeftira, and Benjamin Dillenburger. 2020. "Generative Modelling with Design Constraints: Reinforcement Learning for Object Generation." In RE: Anthropocene, Design in the Age of Humans - Proceedings of the 25th International Conference on Computer-Aided Architectural Design Research in Asia, CAADRIA 2020. Vol. 1.

Amazon. 2023. "Amazon Go." Accessed August 23. https://www.amazon.com/b?ie=UTF8&node=16008589011.

Bennett, Jane. 2010. Vibrant Matter: A Political Ecology of Things. A John Hope Franklin Center Book. Duke University Press.

Bogost, Ian. 2012. Alien Phenomenology, or What It's Like to Be a Thing. Uma Ética Para Quantos? doi:10.1017/CBO9781107415324.004.

Bradski, Gary. 2000. "The OpenCV Library." Dr. Dobb's Journal of Software Tools.

Brayer, Marie-Ange. 2015. "Frank Gehry: The Interlacing of the Material and the Digital." In Frank Gehry, edited by A Lemonier and F Migayrou. Prestel.

Bryant, Levi R. 2011. The Democracy of Objects. New Metaphysics. Ann Arbor: Open Humanities Press.

Campo, Matias del, Alexandra Carlson, and Sandra Manninger. 2021. "Towards Hallucinating Machines - Designing with Computational Vision." International Journal of Architectural Computing 19 (1). doi:10.1177/1478077120963366.

Campo, Matias Del, Sandra Manninger, and Alexandra Carlson. 2019. "Imaginary Plans the Potential of 2D to 2D Style Transfer in Planning Processes." In Ubiquity and Autonomy - Paper Proceedings of the 39th Annual Conference of the Association for Computer Aided Design in Architecture, ACADIA 2019.

Chollet, François, and others. 2015. "Keras." https://keras.io.

Cross, Richard. 2022. "Medieval Theories of Haecceity." In The Stanford Encyclopedia of Philosophy, edited by Edward N Zalta, Spring 2022. Metaphysics Research Lab, Stanford University. https://plato.stanford.edu/entries/medieval-haecceity/.

Davies, E Roy. 2017. Computer Vision: Principles, Algorithms, Applications, Learning. Elsevier Science.

DeLanda, Manuel, and Graham Harman. 2018. The Rise of Realism. Wiley.

Deleuze, G, and F Guattari. 1988. A Thousand Plateaus: Capitalism and Schizophrenia. Athlone Contemporary European Thinkers Series. Athlone Press.

Eisenman, Peter. 2018. The Formal Basis of Modern Architecture. Lars Müller Publishers.

Evans, Robin. 2000. The Projective Cast: Architecture and Its Three Geometries. 1st Paperb. ACLS Humanities E-Book. MIT Press.

Friends of the Rail Park. 2020. "This Is the Rail Park." https://www.therailpark.org/.

Gatys, Leon, Alexander Ecker, and Matthias Bethge. 2016. "A Neural Algorithm of Artistic Style." Journal of Vision 16 (12). doi:10.1167/16.12.326.

Goodfellow, Ian, Jean Pouget-Abadie, Mehdi Mirza, Bing Xu, David Warde-Farley, Sherjil Ozair, Aaron Courville, and Yoshua Bengio. 2020. "Generative Adversarial Networks." Commun. ACM 63 (11). New York, NY, USA: Association for Computing Machinery: 139–44. doi:10.1145/3422622.

Ham, Hanry, Julian Wesley, and Hendra. 2019. "Computer Vision Based 3D Reconstruction : A Review." International Journal of Electrical and Computer Engineering 9 (4). doi:10.11591/ijece.v9i4.pp2394-2402.

Harman, Graham. 2018. Object-Oriented Ontology: A New Theory of Everything. Pelican Books. Penguin Books Limited.

Hosmer, Tyson, Panagiotis Tigas, David Reeves, and Ziming He. 2020. "Spatial Assembly with Self-Play Reinforcement Learning." In Proceedings of the 40th Annual Conference of the Association for Computer Aided Design in Architecture: Distributed Proximities, ACADIA 2020. Vol. 1.

Ingram, David. 2018. "Haecceity and Thisness." Routledge Encyclopedia of Philosophy. Taylor and Francis. doi:10.4324/0123456789-N129-1.

Islam, Arham. 2022. "How Do DALL-E 2, Stable Diffusion, and Midjourney Work?" MarkTechPost. November 14. https://www.marktechpost.com/2022/11/14/how-do-dall%C2%B7e-2-stable-diffusion-and-midjourney-work/.

Karras, Tero, Samuli Laine, and Timo Aila. 2018. "A Style-Based Generator Architecture for Generative Adversarial Networks." CoRR abs/1812.04948. http://arxiv.org/abs/1812.04948.

Klein, Karel. 2020. "Verto Pellis." Offramp Fall (17).

— — —. 2022. "Machines à Rechercher." In Log, edited by Cynthia Davidson. Vol. 55. New York: Anyone Corporation.

Krizhevsky, Alex, Ilya Sutskever, and Geoffrey E. Hinton. 2017. "ImageNet Classification with Deep Convolutional Neural Networks." Communications of the ACM 60 (6). doi:10.1145/3065386.

LeCun, Yann, Léon Bottou, Yoshua Bengio, and Patrick Haffner. 1998. "Gradient-Based Learning Applied to Document Recognition." Proceedings of the IEEE 86 (11). doi:10.1109/5.726791.

Living PlanIT AG. 2021. "Living PlanIT AG." Accessed January 29. http://www.living-planit.com/.

Lynn, Greg. 1999. Animate Form. Princeton Architectural Press.

Mitchell, Melanie. 2019. Artificial Intelligence: A Guide for Thinking Humans. Kindle iOS version. Kindle iOS version.

Morton, Timothy. 2009. Ecology Without Nature: Rethinking Environmental Aesthetics. Harvard University Press.

Morugin, Sergey. 2021. "Ostagram." December 29.

Motzenbecker, Dan, and Kyle Phillips. 2017. "AutoDraw by Google Creative Lab - Experiments with Google." Google Creative Lab. May 6. https://experiments.withgoogle.com/autodraw.

Negroponte, Nicholas. 1972. The Architecture Machine: Toward a More Human Environment. MIT. Press.

———. 1975. Soft Architecture Machines. MIT Press.

Ostagram. 2021. "Ostagram." Accessed January 28. https://www.ostagram.me/about?locale=en.

Park, Taesung, Ming Yu Liu, Ting Chun Wang, and Jun Yan Zhu. 2019. "Semantic Image Synthesis With Spatially-Adaptive Normalization." In 2019 IEEE/CVF Conference on Computer Vision and Pattern Recognition (CVPR), 2332–41. doi:10.1109/CVPR.2019.00244.

Paszke, Adam, Abhishek Chaurasia, Sangpil Kim, and E Culurciello. 2016. "ENet: A Deep Neural Network Architecture for Real-Time Semantic Segmentation." ArXiv abs/1606.02147.

Rehm, Casey. 2019. "'HoaxUrbanism', Deep Architecture - Addin Cui SCIArc 2019 Grad Thesis Prep Magazine by Zhongding Cui - Issuu." SCIArc 2019 Grad Thesis Prep Magazine. https://issuu.com/addincui/docs/addin_20190406_magzine_spread_100dp.

Rowe, Collin, and Fred Koetter. 1983. Collage City. MIT Press.

Schulze, Franz, and Edward Windhorst. 1985. Mies van der Rohe: A Critical Biography. University of Chicago Press.

Soltani, Amir Arsalan, Haibin Huang, Jiajun Wu, Tejas D. Kulkarni, and Joshua B. Tenenbaum. 2017. "Synthesizing 3D Shapes via Modeling Multi-View Depth Maps and Silhouettes with Deep Generative Networks." In Proceedings - 30th IEEE Conference on Computer Vision and Pattern Recognition, CVPR 2017. Vol. 2017-January. doi:10.1109/CVPR.2017.269.

Stuart-Smith, Robert, and Patrick Danahy. 2022a. "Visual Character Analysis Within Algorithmic Design, Quantifying Aesthetics Relative to Structural and Geometric Design Criteria." In CAADRIA 2022, POST-CARBON - Proceedings of the 27th CAADRIA Conference - Vol. 1, Sydney, 9-15 April 2022, edited by Jeroen van Ameijde, Nicole Gardner, Kyung Hoon Hyun, Dan Luo, and Urvi Sheth, 131–40. Sydney. doi:10.52842/conf.caadria.2022.1.131.

———. 2022b. "3D Generative Design for Non-Experts: Multiview Perceptual Similarity with Agent-Based Reinforcement Learning." In Critical Appropriations - Proceedings of the XXVI Conference of the Iberoamerican Society of Digital Graphics (SIGraDi 2022), edited by PC Herrera, C Dreifuss-Serrano, P Gómez, and LF Arris-Calderon, 115–26. Lima: SIGraDi.

Tish, Daniel, Nathan King, and Nicholas Cote. 2020. "Highly Accessible Platform Technologies for Vision-Guided, Closed-Loop Robotic Assembly of Unitized Enclosure Systems." Construction Robotics 4 (1–2). doi:10.1007/s41693-020-00030-z.

Venturi, Robert. 1977. Complexity and Contradiction in Architecture. Museum of Modern Art.

Wang, Zhou, Alan Conrad Bovik, Hamid Rahim Sheikh, and Eero P. Simoncelli. 2004. "Image Quality Assessment: From Error Visibility to Structural Similarity." IEEE Transactions on Image Processing 13 (4). doi:10.1109/TIP.2003.819861.

Wolfe, C. 2010. What Is Posthumanism? Posthumanities Series. University of Minnesota Press.

Wu, Jiajun, Chengkai Zhang, Tianfan Xue, William T. Freeman, and Joshua B. Tenenbaum. 2016. "Learning a Probabilistic Latent Space of Object Shapes via 3D Generative-Adversarial Modeling." In Advances in Neural Information Processing Systems.

Yousif, Shermeen, and Daniel Bolojan. 2021. "Deep-Performance: Incorporating Deep Learning for Automating Building Performance Simulation in Generative Systems." In Projections - Proceedings of the 26th International Conference of the Association for Computer-Aided Architectural Design Research in Asia, CAADRIA 2021. Vol. 1.

◀ ◀

8.1 Viscera/L
*Close-up of semi-
autonomous clay
sculpting outcome.
An articulated robot
determined the carving
sequence in response
to sensor feedback,
continuously reacting
to the current state of
the clay body.*

8 | Adaptive Manufacturing
Towards Situated, Embodied Production

Design as an Embodied Act

Picture this: Paint is flung across the room, landing in streaks and splats onto a canvas far greater in size than Jackson Pollock himself, who is gesturing wildly throughout the process. In the art world, it is not unusual for highly sought-after work to be created through the employment of strange or lively bodily acts. In the professional spheres of architecture or product design, work is rarely created this way. While design is seldom a linear process, it is closely associated with analytical methods that inform activities heavily reliant on premeditation and human visual cognition. Intuitive it can be, yet intuition is largely reserved to graphics-based work, with little attention given to the physical body's role in such processes.

Pollock's *'Drip Paintings'* and other work (Harrison 2014; Landau 2000) are not solely determined through visual cognitive processes, there is a clear causative relationship between the motion of Pollock's own body and the splattering of paint lying across his canvases, where a series of cascading material and physical effects were caused by his actions, and in turn, impacted his subsequent actions. Similarly, Henri Michaux generated a range of gestural and field-based graphical works in ink and paint through what he described as the conscious, chaotic motion of his body. Michaux's *'Mescaline Drawings'* were produced by speed, repetition, and Brownian motion under the influence of the drug Mescaline which altered his perception and consciousness (Michaux et al. 2000, 131). Michaux exploited both cognitive and bodily motion potentials to create visceral work.

To architectural examples, Frank Gehry's production of scaled models during a building's concept design stage is heavily influenced by his physical manipulation of paper, cardboard and other materials, whose formal character he seeks to preserve in final built projects, while their physical properties constrain a design's geometry to developable surfaces suitably rationalised for production (Brayer 2015, 173). CoopHimmelblau's Wolf Prix and Helmut Skisinki's initial concept for *Open House* commenced with a hand sketch drawn eyes-closed, whose gestural qualities they sought to preserve throughout the entire design and construction process (Werner and CoopHimmelblau 2000, 176). Their design for Melun-Sénart and other masterplans was conceived in brainstorming sessions that leveraged hand signs and body gestures more than verbal exchanges (Werner and CoopHimmelblau 2000, 177–180). These activities demonstrate an architectural interest to go beyond total reliance on the architect's visual cognition, to incorporate gesture, action, and effect within the creative design process.

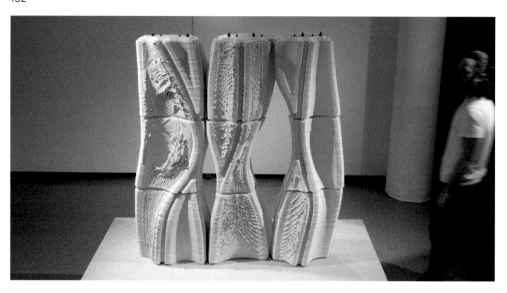

▲

8.2 Robotic Prometheus
*Multi-part assembled
additively manufactured
ceramic prototype.*

Where these architectural examples may seem fanciful due to a distinctive separation between the acts of the designer's bodies and the physical media of their buildings, this was not the case in Michaux's or Pollock's work. If, however, such behavioural acts are directly related to the physical matter of architecture itself, a more considered, technically proficient design outcome might be achievable that employs more than human visual cognition to achieve its goals. Our bodies do not depend solely on visual cognition or premeditated action, our other senses play a part also, as does proprioception – our perception of stimuli relating to internal conditions such as position, posture, or equilibrium (Encyclopaedia Britannica 2021). Proprioception plays an instrumental role in learning complex physical activities such as kung fu, piano or ballet. If we already engage these faculties for creative purposes, how could such capabilities be harnessed for architectural design activities when our bodies are limited in scale, strength and ability to manipulate architectural materials? Perhaps vicariously through other bodies?

Behavioural Intelligence

In 1953, while NASA still employed human *"computers"* to calculate the orbit of their rockets (Holland 2018), and more than 20 years before the PC (personal computer) industry took off (Encyclopaedia Britannica 2022), Walter Grey Walter developed three-wheeled robot vehicles that could autonomously navigate a room. This was no small feat, the robots were able to wander around, seek and avoid obstacles and walls, and return to a charging station when their battery was low, only to continue explorations when fully charged (Walter 1953). The two *Machina Speculatrix* vehicles, commonly referred to as the *"Tortoises"*, were simple analogue machines that carried two sensors, two actuators (for motion and steering), a light, and battery (Walter 1953, 126–30). In the tortoises, Walter envisaged an experiment that could demonstrate intelligence through situated, autonomous behaviour (Walter 1953, 120). The principles used to guide the tortoises' programming operated through simple event-based rules that allowed the tortoises to appear to have a will of their own (Arkin 1998, 9), exploring spaces, moving towards light sources and away from obstacles, and exhibiting unpredictable behaviours such as dancing in front of mirrors and avoiding or clustering with other tortoises (Walter 1953, 179). In Walter's book *The Living Brain*, he cites both Norbert Weiner's Cybernetics and Clerk Maxwell's paper on Watt's steam engine governor as sources of inspiration (Walter 1953, 26).

Conceived in the late 1940s by Norbert Wiener, Cybernetics was a form of systems theory that attempted to describe the control and communication of both machines and living organisms (Wiener 2019). In coining the word *'Cybernetics'*, derived from the Greek word for governor "κυβερνήτης" (Wiener and von Neumann 1949), Wiener was paying homage to Clerk Maxwell who argued that Watt's governor was a significant development, enabling the self-regulation of pressure in steam engines (Wiener and von Neumann 1949; Wiener 2019, 97). The governor's principle of negative feedback that Wiener admonished is widely used today, such as in air-conditioners where thermostats enable the self-regulation of temperature. Walter considered Watt's governor to be the first operational model of a reflex circuit such as a nerve or muscle, and the distinguishing principle that enabled his tortoises' use of sensor-data to produce unpredictable behaviour compared to historical automata (see Chapter One) that could only perform predetermined actions (Walter 1953, 115–16).

For Wiener, cybernetic principles of self-regulation and feedback enabled diverse systems to adapt to live circumstances, where contradictory conditional rules can provide a negotiated, self-regulated solution. Such event-based programming can be as simple as:

If this happens, do that.

or

▼ *If having done that, X occurs, do the first thing again or something else.*

8.3 Robotic Prometheus
Close-up photographs of ceramic prototype after glaze firing.

In Walter's tortoises, such rules are defined before the action begins, then acted out through a series of events and interactions over time,

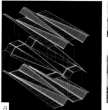

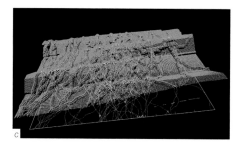

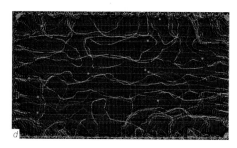

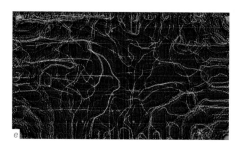

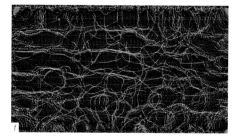

characteristically predictable while variable in actual behaviour due to outside perturbation giving rise to chaotic divergences.

In Artsbot-2003 and similar works, Artist Leonel Moura utilises a similar robot platform to Walter to produce art. Robots are programmed to explore a canvas and draw with markers where they perceive existing marker lines. In these works, several robots collectively produce the art, reacting to each other's previous actions. While these robots operate autonomously, Moura describes the work as symbiotically created between the robot drawers, the artist and collaborators who develop the robot and the program that governs each robot's behaviour (Moura 2013). Both the tortoises and Artsbot-2003 robots exhibit locally adaptive behaviours simply by *reacting* to their environment. This approach aligns with a sub-field of robotics commonly referred to as *reactive robotics* where a robot operates in direct response to a perceived event condition.

Behaviour-Based Control
In scientific robotics research, there are several competing approaches to robot control. In *The Robotics Primer*, roboticist Maja Matarić describes these through simple statements, equating *Reactive Systems* to *"don't think react"*, *Deliberative* as *"think hard, act later"*, *Hybrid* as *"think and act separately in parallel"* and *Behaviour-Based control* as *"think the way you act"* (Matarić, Koenig, and Arkin 2007). Matarić critiques each approach, suggesting reactive systems are inflexible and incapable of adaptation or learning, deliberative systems are too slow, and hybrid systems – too complex (too error-prone and insufficiently robust). In contrast, Matarić advocates for a *behaviour-based* approach that builds complex behaviour through the assemblage of simple, consistent components (Matarić, Koenig, and Arkin 2007, 188).

An extension of reactive control, behaviour-based control was first proposed by roboticist Rodney Brooks, the former head of MIT's Computer Science and Artificial Intelligence Laboratory (CSAIL) and a founder of robot vacuum-cleaning company, iRobot™. Brooks was sceptical of computationally intensive deliberative approaches to robot control that seemed to produce less situational intelligence than ants or termites exhibit despite their very simple neurological

◀

8.4 situatedFabrications
a) 3D model of formwork layers to be incrementally added throughout manufacturing, b) an algorithm guides the extrusion nozzle to move closely over the top of an already deposited layer, c) multi-agent design-fabrication simulation with point cloud of last 3D scan heuristically adjusted by more recent material depostion, d-f) time-frames from multi-agent design-fabrication simulation. Each trajectory represents a robot deposition toolpath.

systems. Like Walter, Brooks considered intelligence to be something sensed rather than computed or simulated. Brooks insisted that to engage with the physical world, robotic approaches should be:

- *Embodied:* experiencing the world within a physical body of robot hardware, and,

- *Situated:* embedded in the world, sensing and acting in real-time in relation to an environment.

To illustrate this, Brooks describes an airline reservation system as situated but not embodied, and an automated spray-painting industrial robot as embodied but not situated (Brooks 2002, 51–52). To engage with the physical world, this chapter explores design approaches that are *embodied* and *situated*, leveraging semi-autonomous design decisions in real-time within a behaviour-based approach to manufacturing.

▼

8.5 Control Approaches
A small sample of robot control examples from this chapter: a) Automated Articulated Robot, b) Artitculated robot with real-time feedback, c) Aerial robot with external localisation from motion capture cameras, d) Fully autonomous, using on-board sensor feedback to inform decisions on how to move through sequences of motion commands. Option (d) does not require but can calculate localisatoin in cartesion coordinates.

Brooks proposed an approach he called *'Subsumption Architecture'* which conceptually simplified robot control down to a process of priority-based arbitration amongst a series of parallel executed simple behaviours that could collectively produce more complex actions. The approach enabled a programmer to start with simple behaviours and incrementally add more elaborate sets of rules over time (Arkin 1998, 135). Brooks' Subsumption Architecture attempts to develop a robot's actions from the interaction of different behavioural rules that are reactive to both the environment (via sensor data), and concurrent internal states or actions of the robot. In Brooks' legged robot Ghengis, for instance, walking is not explicitly programmed, but achieved by rules that relate the actions of one leg to that of others (Brooks 1985). Although Brooks advocated for a tight integration of hardware and software that ties low-level control of a robot's body plan to its sensor and actuator systems, the work presented in this chapter is focused on high-level motion planning applications that are often built on top of lower-level control systems[1] (largely due to a focus on design and fabrication applications rather than robotics engineering). In some of this work, behaviour-based programming is extended using

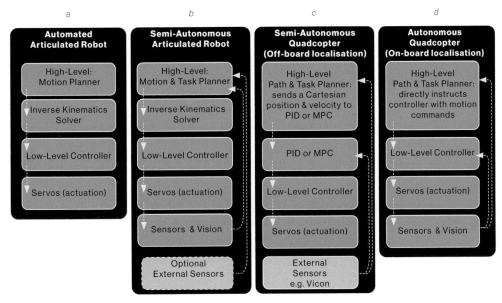

8.6 situatedFabrications
*a-c) Manufacturing is
adapted to move around
successively added
void-formers, d)
resulting concrete panel,
additively manufactured
by an ABB IRB2600 robot.*

computer vision for localisation[2] and sensor-feedback, and therefore might be considered to operate as a Hybrid control approach, which Matarić describes as being able to *"think and act separately in parallel"* (Matarić, Koenig, and Arkin 2007, 188).

In earlier chapters, agent-based modelling offered a generative design approach that was developed through the incremental addition of behaviour-based algorithmic rule sets. Emergent behaviour arose from the interaction of simpler behavioural building blocks in a similar manner to Brooks' Subsumption Architecture. This same approach can be applied to high-level motion planning on diverse robot platforms where agent behaviour operates within the constraints of a robot's body plan and low-level control system to adapt to spatiotemporal events to engage in manufacturing activities. Behaviour-based programming offers a systemic approach to design and manufacturing that can tie robot control to computational design approaches. In this chapter, behaviour-based manufacturing extends generative design into physical materials and environments. By employing computer vision and other sensing technologies, and operating within the limits of a robot's physical capabilities and end-effector manipulators, design operates through embodied fabrication processes.

Design via Robot Behaviour

The robotic manipulation of matter is one area in which a creative design process can be embodied within a robotic fabrication task. In the University of Pennsylvania's Masters of Science in Design: Robotics and Autonomous Systems program (MSD-RAS)[3], students were provided with a custom-developed agent-based program and tasked with employing it to support the robotic manipulation of clay through carving or smooshing operations (where an agent's motion trajectory is translated into a robot tool-path)[4]. One thesis project, *Robotic Prometheus*, applied this process to the smooshing of additively manufactured ceramic parts prior to their bisque firing while the clay was still malleable. As the material outcome would have been difficult to simulate, the design process partially took place in pre-production coding and partly through the development of robotic smooshing gestures in physical prototypes, created by pressing an incremental metal forming (IMF) tool with a ball-point tip into the surface of manufactured clay parts (Figures 8.2, 8.3). In a Brooks behaviour-based programming sense, this coupling of a robot's physical actions with its algorithmic programming offers an "embodied" design approach, that provides material engagement via the robot's physical body motion and its end-effector tool to produce unique aesthetic design outcomes.

As no feedback was incorporated into *Robotic Prometheus's* fabrication process, it might be considered to be a deliberative approach to robot planning. This is typically the best approach to manufacturing, as one aims to reduce the complexity and risk of operations by eliminating unknowns. Industrial robot arms (articulated robots) are typically fixed in position, with materials also placed in known locations to enable manufacturing to leverage offline (predetermined) programming to produce predictable outcomes. However, there are scenarios in which this is not adequate;

1. When the environment a robot is operating within is dynamically changing,

2. Where a robot's tasks are constantly changing while programming remains unchanged, or,

3. When the materials a robot is engaging with are dynamically changing or imprecisely located.

▼

8.7 situatedFabrications
Concrete additively manufactured prototype. a-b) Views of one of two final prototypes with inner cavitities visible thorugh voids on the exterior.

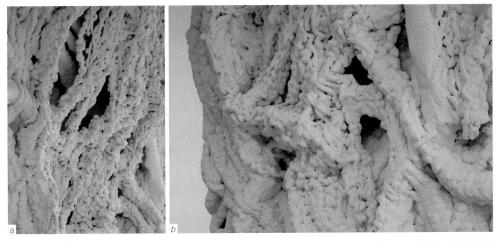

a   b

In relation to (3), material behaviour in *Robotic Prometheus* was unpredictable to a degree; however, it was predictable enough to be successfully executed without any sensor feedback. The approach essentially operated blindly, with the robot following a predetermined motion sequence, limiting its ability to engage with the material in variable ways. If the material fabrication process was more unpredictable, sensing and feedback would be required to ensure success, rendering the approach as also "situated".

In a guest professorship studio, *Situated Fabrications* at the University of Innsbruck[5], two multi-layered concrete panels were additively manufactured by an industrial robot arm. A series of voids and cavities were created in the panels through the intermittent manual placement of void formers throughout manufacturing (Stuart-Smith 2018). This additional complexity created a degree of unpredictability in the manufacturing process, necessitating both scanning-in-the-loop and heuristic programming to accommodate variable changes in manufacturing depth produced from the frequent placement of void former parts and the variable viscous deposition of concrete (discussed in Chapter Six) (Figures 8.4, 8.6). The method enabled manufacturing to closely follow the topographic condition of already fabricated material with a specified offset, without the robot colliding with already manufactured material (MEX-AM manufacturing is hard to control without maintaining a deposition offset distance).

Concrete was extruded continuously from a simple robot end-effector tool whose six-axis motion tool path was determined by a custom multi-agent program. The approach was conceived of as an iterative layering process, where a population of agents was tasked with moving across a bounded simulation space in 2D where their trajectories were projected onto a periodically updated 3D scanned environment obtained using a Microsoft Kinect™ RGBD camera attached to the robot arm.

Each agents' vertical position was continuously adjusted in relation to the scan data, vertically offset above the scan data by a value equivalent to the concrete's material deposition thickness. Agent trajectories were then utilised as a robot extrusion tool path. To

8.8 Viscera/L
Custom robot end-of-arm tool profiled for custom sculpting. Includes RGBD camera and laser distance sensor.

reduce scanning interruptions, a single industrial robot manufactured multiple discrete agent trajectories between each scan and applied a heuristic to estimate the current state of manufacturing between scans. An agent crossing a previously manufactured agent trajectory (post-scan) is obliged to move up and over it to avoid the robot tool colliding with recently deposited material. To integrate this in the software, the creation of an agent trajectory causes a local shift upwards in the Z-axis to parts of the last-obtained 3D scan data equivalent to the material extrusion height. These modifications remain in place until overwritten by a new 3D-scanned topography (Figure 8.4).

8.9 Viscera/L
a-b) Custom robot end-of-arm tool, c) additive manufacturing, d) subtractive manufacturing of (c) using custom tool, e-f) top view and elevation of a second part prior to subtractive manfuacturing, g) gestural effects created by the profiling of the custom tool.

Although different in material organisation, the application of the same algorithmic approach (with slight variations in agent behaviour-based rules) for the two panels successfully resulted in a coherently authored material complexity (Figure 8.7). The unpredictability of the concrete's viscosity during deposition and the periodic placement of void formers necessitated the incorporation of scanning-in-the-loop within design, rendering the process as *situated*. The software's input into iterative offline industrial robot programming also ensured the design approach was *embodied*. Critically, however, the algorithm was slightly de-coupled from the robot's body and perception faculties due to the fact that agent motion did not adapt to 3D scan data and was only re-mapped in the Z-axis. If autonomous perception is used to adapt agent behaviour, a more situated form of design-fabrication is possible that is more suited to the automated manufacturing of bespoke design works.

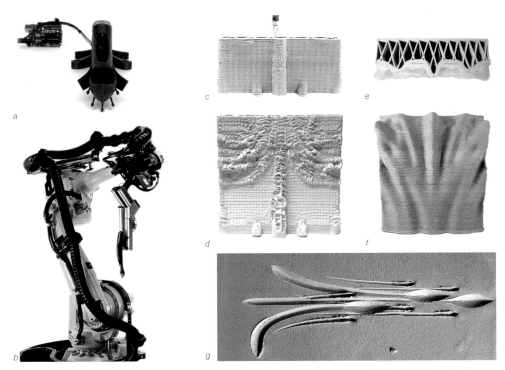

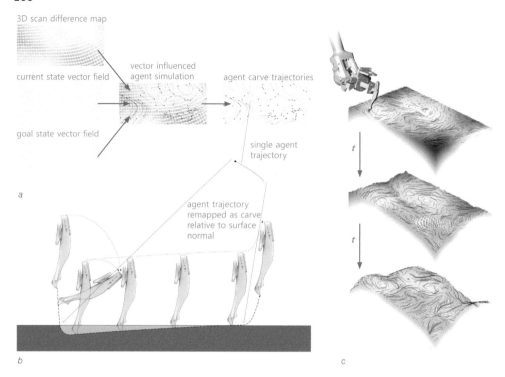

3D scan difference map

current state vector field

vector influenced
agent simulation

agent carve trajectories

goal state vector field

single agent
trajectory

a

agent trajectory
remapped as carve
relative to surface
normal

b

c

8.10 Viscera/L
a) sculpting trajectory
generation, b)trajectory
sculpting operation, c)
sculpting sequence.

▲ **Adaptive Manufacturing**

Where *Situated Fabrications* augments the industrial process of concrete additive manufacturing, the research project *Viscera/L*, creatively extends the artisanal process of ceramic sculpting and is intended to operate within a combined Additive and Subtractive Manufacturing approach (ASM) (Figure 8.9). Sculpting requires an artist to sensitively engage with a constantly changing material condition to arrive at an intended design outcome. As discussed in Chapter Three, this creative process is not completely predetermined but is also influenced by a continual process of material engagement where the sculptor adapts to the clay's affordances (Malafouris 2008). In the building industry, sculpting has been largely replaced by CNC subtractive manufacturing which utilises an offline (predetermined) program to undertake an automated carving process that is executed without sensing or feedback, to produce a monolithic object devoid of machinic expression. While substantial research has been done into the roboticisation of clay sculpting on industrial robots (Tan and Dritsas 2016; Ma et al. 2020), this has primarily involved standard sculpting tools and offline tool path programming for explicitly modelled or computationally derived geometries without feedback from temporal material states. In contrast, *Viscera/L* explores a real-time semi-autonomous approach to incremental sculpting intended to produce specific design character (Stuart-Smith et al. 2022).

Developed in the Autonomous Manufacturing Lab at the University of Pennsylvania (AML-Penn), the method is both embodied and situated. Embodiment goes beyond the physical industrial robot platform to incorporate a bespoke robot end-of-arm tool (EOAT) whose physical profiling is designed to develop gestural effects during sculpting operations, which required the tool geometry to be iteratively developed alongside a series of carving experiments. The tool also incorporates an Intel Realsense™ RGBD camera and a laser sensor to support intermittent computer vision and 3D scanning together with more frequent position-correction feedback throughout manufacturing (Figures 8.8, 8.12). The algorithmic design approach is situated by enabling sculpting operations to be dynamically determined in relation to tool sensor data feedback (Figures 8.10, 8.11).

A robot sculpting path is developed through agent-based motion in a similar manner to other projects discussed earlier in this book; however, agent behaviour in *Visceral/L* seeks to address differences between an autonomously perceived current state, and a specified goal state (a 3D model) (Figure 8.11). Throughout manufacturing, machine vision and semi-autonomous motion planning are used to adapt sculpting tasks to areas with significant deviation from intended geometric results. Periodically obtained 3D scan and computer vision data of the current state are compared to the desired geometric outcome. The differences between these two states' topography and curvature are translated into a single 3D vector field that operates as an environmental influence in a multi-agent simulation. Agents move in relation to the vector field and agent-to-agent behavioural parameters that correspond to tool-scale sculpting operations, to produce a group of sculpting trajectory paths. This process is repeated regularly to form part of a multi-step iterative sculpting process that transforms a clay part from its initial material condition down to its desired final state (Figure 8.11). In contrast to the previous projects, agent-generated tool paths operate with sensor feedback (using an online programming strategy determined during the robot's operation), sent to an ABB industrial robot at high-frequency using the *Robot ExMachina* framework (del Castillo y López 2019).

▼
8.11 Viscera/L
*2-Axis time sequence of:
a) initial clay conditions,
b) scan and vector data,
c) agent trajectories, d)
sculpted result.*

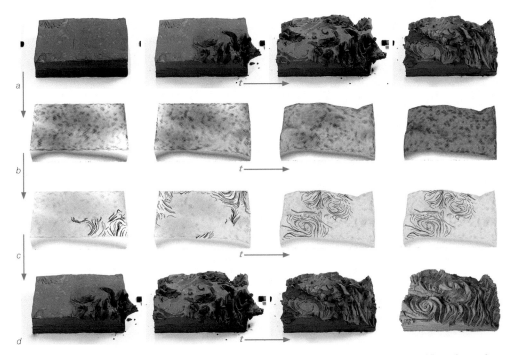

a

b

c

d

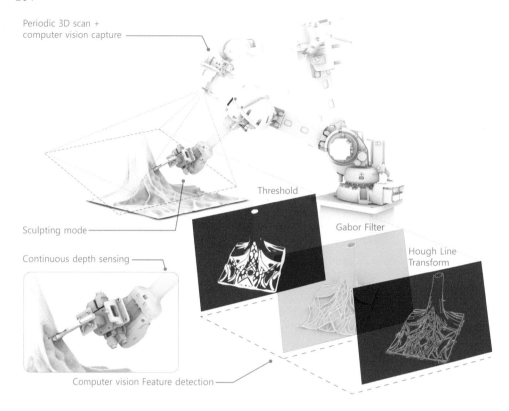

Periodic 3D scan +
computer vision capture

Sculpting mode

Continuous depth sensing

Threshold

Gabor Filter

Hough Line
Transform

Computer vision Feature detection

◀◀
8.12 Viscera/L
*Top and aerial views
of two differentiated
outcomes (a & b, c
& d) that shared an
identical goal state
but commenced with
different initial clay states.*

▲
8.13 Viscera/L
*Autonomous perception
is used to provide
continuous laser sensor
data and periodic
RGBD scanning and
computer vision feedback
throughout manufacturing.*

To test the method, two thick clay slabs with different initial volumes were sculpted to achieve the same desired geometric outcome. As expected, the method successfully sculpted the differently shaped blocks of clay into two similar but unique outcomes that were alike in volume and shape, but different in their organisation of sculpted material expression (Figure 8.12). Differences in outcome arose not only due to the dissimilarity of their initial material volumes, but also the impact this had on a divergent set of sensing and adaptive gestural actions. The physical properties of the tool, together with sensing and event-based rules contribute to the emergent outcomes. Not entirely predictable, the outcomes are an example of Chaos Theory where a small change in initial conditions can give rise to large differences in a final outcome – what Edward Lorenz terms as "sensitive dependence" (Lorenz and Hilborn 1995).

Posthuman Material Engagement
Digital architecture is typically developed within 3D graphics applications and in corporeal algorithms. Designs might also be iteratively honed using human visual assessment and feedback before potentially being executed by robotic fabrication routines. In contrast, *Viscera/L* explores non-human visual cognition to partly author creative and manufacturing action. This Posthuman condition does not remove the human designer but places them in a strange yet intimate relationship with non-human sensation. This almost synaesthetic condition for the human designer operates through spatiotemporal event-based gestures rather than

visual media. In *Viscera/L*, design is orchestrated by, and realised vicariously through, the designers' dialogue with software, hardware and material. Design agency operates within all party's contributions, with each influence registered within the design outcome in different ways.

Similar to art historian Heinrich Wölfflin's assessment of Antonio Bernini's Roman Baroque sculptures offering a departure from describing form through continuous contours and compositional planes to become *"broken up with 'accidental' effects to give them greater vitality"* (Wölfflin, Murray, and Simon 1966, 35), the *Viscera/L* prototypes arise from a multitude of small partial carving motions. These ornamental-like qualities are not semantic, but derived from embodied visceral acts. Arguably, this provides the work with an alternative form of vitality, directly informed by the sensing and robotic manipulation of matter. Further, this bespoke design character is maintained across the two manufactured outcomes despite their geometric differences. For a bespoke approach to hold promise for actual architectural applications, however, it must be scalable to industrial volumes of production. Theoretically, *Viscera/L's* method achieves this by automating the production of the unique through semi-autonomous perception and decision-making within a combined approach to design and manufacturing that employs real-time robot motion planning, computer vision, sensor feedback, and bespoke tooling. But in practice, robotic manufacturing in the building industry does not currently support bespoke works en masse. A solution, however, is in the works.

▼

8.14 Viscera/L
Work-in-progress on an ASM column head part: a) 3D scan, computer vision identification of: b) deep pockets, c) edge features, d) initial clay body, e) edge vectors, f) agent sculpting trajectories. Manufacturing process: g,h) additive manufacturing, i) subtractive manufacturing.

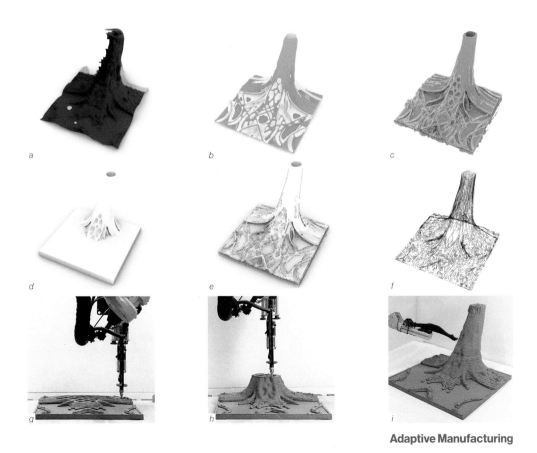

a b c

d e f

g h i

Adaptive Manufacturing

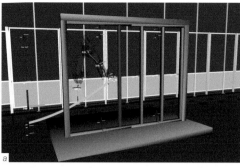

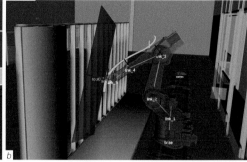

a

b

a

b

▲
▲
▲
8.15 M-RAM
*a & b) visualisations
from simulation. Blue
motion plan updated in
real-time while robot
avoids collisions with
already built parts.*

▲
8.16 M-RAM
*a) Collaborative assembly
visualisation, b) AML-
UCL lab space: mobile
Robotniq with UR10 arm
and Kuka track robot.*

Adaptive Multi-Robot Manufacturing

In current building industry manufacturing approaches, non-standard geometric variation usually necessitates additional labour, time, material waste, risk and ultimately, cost. While individual building parts can be fabricated to variable dimensions, to date there are limited approaches to automating variation in more general building assembly activities (off-site or on-site). To move past established methods; however, is challenging given that the industry's present productivity levels are insufficient. In the UK, current operations cannot be adequately scaled to meet the demands of larger-sized project logistics (see Chapter Two[6]). Studies have concluded that productivity can be improved by 60% simply by undertaking at least 70% of construction activities off-site (Science and Technology Select Committee 2018). Modular construction offers one means to do this, involving the off-site assembly of large volumetric building modules that are relatively quickly stacked like Lego™ on a construction site, minimising on-site construction activities. Two of modular's biggest proponents though, Laing O'Rourke™ and Katerra™, have faced several challenges in operating without automated solutions to assembly[7]. There is still a large gap between commercial companies' manual assembly approaches (Gordon 2023) and research taking place in institutions such as ETHZ, whose DFab House in the Empa Nest building demonstrates the robotic assembly of non-standard dimensioned prefabricated timber-frame building modules (Willmann et al. 2016).

There are also some fundamental differences between construction and other forms of manufacturing that pose several challenges to automating building assembly. While factories typically separate workers, machinery and the object being manufactured into separate zones, the large scale of buildings and prefabricated building modules make such separations impossible. In both on-site and off-site building assembly, the scale of the work necessitates workers operating inside of the space they are modifying. Building work involves hazardous tasks, such as working at height, lifting or working overhead, or next to heavy machinery which can lead to injuries and fatalities (see Chapter Two). Even ETHZ's novel DFab House involved people working inside the manufacturing work cell adjacent to large robot systems (Willmann et al. 2016). An alternative, semi-autonomous robotics solution to building assembly could alleviate the need for workers to enter the manufacturing zone, allowing them to focus on more safe activities, whilst increasing productivity and possibilities for automating non-standard geometrically varied assembly.

In research at UCL's Autonomous Manufacturing Lab (AML-UCL), a series of discussions with construction companies Skanska and Mace led to feedback that robots able to perform multiple assembly tasks would be more useful than task-specific robots. This is fundamentally different to prevalent manufacturing industry methods, where robots are primarily used to repeatedly manufacture the same part/s in a fixed work cell, with task-specific jigs and tooling, instructed by offline programming (*embodied* but not *situated*). Current practices though, are disadvantageous for on-site and off-site building assembly activities that encompass substantial dimensional or geometric variation, rather than a single manufactured outcome. To achieve flexibility, however, is incredibly complex and requires robots to be both *embodied* and *situated*. As robots assemble building elements, their previous building activities create further obstacles that robots must avoid and operate around. To physically move and assemble elements, a multi-axis robot such as an industrial robot arm must determine a kinematics solution that enables motion within the constraints of each of its joint's rotations whilst avoiding collisions with its own body, payload or the environment. Given the immense scale of assembly tasks, mobile robots are appropriate for some applications in both on-site and off-site work, while fixed robot systems might also need to move considerable distances (e.g., on a track). With increased mobility comes greater environmental uncertainty, and the need for more degrees of autonomy.

▼
8.17 Husky+
a) Husky+ platform with lidar and 3D scanner, b) testing on a live construction site.

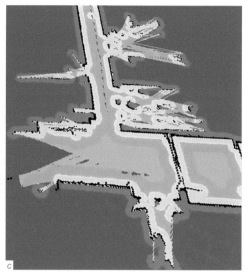

▲

8.18 Husky+
a) 360deg image
capture, b) 3D scan
data capture, c)
frontier mapping of
autonomously perceived
environment.

As a robot navigates within the build work zone, it needs to re-calculate a feasible motion plan that avoids obstacles, some of which are dynamic such as other robots or machinery, requiring adjustments to its planned course of action. This necessitates "online" motion planning, where the robot can continuously adjust its path to provide robust error-free task execution.

A UK EPSRC/Transforming Construction Challenge funded project (EPSRC EP/S031464/1 2018) being developed in the AML-UCL, is focused on the development of a multi-robot adaptive manufacturing (M-RAM) software framework to support semi-autonomous off-site assembly operations (Figure 8.15). M-RAM will utilise industry-available fixed, track, and gantry-based industrial robot arms alongside custom-developed mobile robot platforms (EPSRC EP/S031464/1 2018). The manufacturing approach is design-agnostic and aims to provide a flexible, shared robot work cell where a heterogeneous team of robots can adaptively assemble building elements from a 3D Building Information Model (BIM). The 3D BIM file will act as a real-time digital twin (see Chapter Two), keeping track of the current state of building activity. To achieve this, multi-agent programming provides a cognitive layer to multi-robot manufacturing, enabling individual and collective building activities to be undertaken within a self-organising, adaptive system. The framework is to be installed as a high-level planning software on each robot, providing capabilities for robots to undertake autonomous task determination, path planning, mobile navigation, kinematic motion-planning, and multi-robot task sequencing and optimisation. It will leverage online robot programming and support high-precision manufacturing using in-the-loop metrology and will initially be demonstrated using a narrow set of assembly activities.

The M-RAM framework will first be tested on a simple prototypical wall system assembly (Figure 8.16). In response to a 3D BIM model, robots will autonomously determine the tasks and sequence of operations they need to undertake relative to their own spatiotemporal constraints, decomposing tasks into smaller event-based decisions. This behavioural agent-based approach enables live-adaptive assembly sequencing to be strategised through the orchestration of events. Several challenges must still be overcome to demonstrate M-RAM. If successful, the approach will serve as a proof of concept for multiple robots assembling in parallel off-site that could potentially be implemented to operate around the clock (24/7) with minimal supervision. Developing similar capabilities for on-site construction, however, is beyond the scope of AML-UCL's current funding. Although, the lab is undertaking research into the use of mobile robots for on-site data collection.

Autonomous On-site Data Collection

On a Mace construction site in central London, a human contractor visits the site once a week to do a walk-through, passing across each floor with a 360 degree camera strapped to a pole rising a half-metre above the contractor's safety hat. This visual data is fed into a shared project database where it is correlated with the architect's floor plans and allows the general contractor (builder), architects and consultants to track and review building progress, identifying errors and imperfections as they arise, and in some cases preventing their occurrence in the first place. If more accurate three-dimensional information is required, a surveyor visits the site to do a partial 3D scan using manually operated LiDAR equipment. Considering the fast pace of work and the complex coordination of several sub-contractor parties' activities, current approaches fail to capture a complete picture of progress or keep a sufficient record of as-built information. Design and construction oversight activities are still reliant on frequent in-person site visitation. There are also unfortunate circumstances where construction companies must remove completed parts of the build to inspect poorly or incorrectly built work that is not visible during later stages of construction. Delays and re-work from these activities can generate losses in millions for a construction company, where a chronologically layered as-built digital-twin BIM model might be able to avoid or help remedy such catastrophes more rapidly. In partnership with Mace, the AML-UCL is developing an autonomous roaming robot to address these issues, capable of providing rapid, high-frequency on-site data collection.

▼
8.19 Quadcopter Motion
a) Perspective view of quadcopter. Motion is achieved by different weightings of 4 propellors: b) pitch, c) roll, d) thrust, and e) yaw.

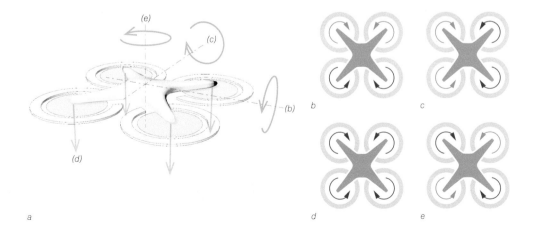

a

b   c

d   e

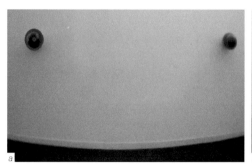

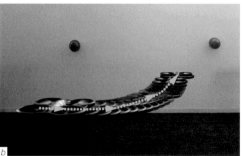

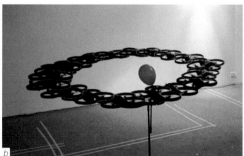

8.20 Autonomous Behaviour
Quadcopter object recognition of a balloon.

8.21 Autonomous Behaviour
Quadcopter seeks red balloon while avoiding blue balloon.

8.22 Autonomous Behaviour
Quadcopter circles blue balloon at a distance.

8.23 Autonomous Behaviour
Line-following. Quadcopter moves to maintain view of red line in the centre of its down-facing camera.

An AML-UCL compiled software framework drives *Husky+*, a ClearPath Robotics™ 4-wheeled Husky™ robot platform (Clearpath Robotics Inc. 2023) coupled with a LiDAR and 360 degree camera to undertake autonomous navigation, and on-site data collection (Figure 8.17). As the construction site environment is constantly changing, unknown static and dynamic obstacles make robust offline pre-programming of *Husky+'s* movements impossible. Instead, *Husky+* must navigate autonomously, while attempting to optimise the location and quantity of data samples it collects. Using a second 2D LiDAR, a method known as Frontier Mapping is used to develop a 2D planimetric map of its surrounding environment and rules that encourage exploration of unmapped regions (Butters et al. 2019), while its knowledge of past experiences enables it to navigate back to a location where its battery supply can be replenished. *Husky+* avoids people, identified using computer vision and machine learning methods, while it captures 360 degree photos and 3D LiDAR scans at a series of self-identified locations throughout a journey that it is capable of undertaking several times a day. In trials, *Husky+* was able to autonomously upload collected data to a remote server, creating the possibility for human collaborators engaged in the project to have up-to-date and high-frequency historical information on building progress (Figure 8.18). This would not be possible without *Husky+'s* ability to perceive and navigate around its environment autonomously. Although *Husky+* is programmed to perform a pragmatic task, similar approaches can also be leveraged creatively.

Towards Situated and Embodied Autonomous Design

From 2012 to 2017, Studio Stuart-Smith in the Architectural Association's (AA) Design Research Laboratory (S-S_DRL), leveraged approaches to autonomous robot motion to support creative design applications. As described in the next chapter, these primarily focused on the use of quadcopter aerial robots. An aerial robot embodies different degrees of freedom (DOF) to other robot platforms, resulting in different behavioural affordances. Quadcopters consist of four propellors, arranged in an X-configuration with one set of diagonally opposite propellors typically spinning in one direction and the other diagonal pair rotating in the opposite. Control of the propellor's motors is abstracted to four coordinated types of motion commonly used in aerial vehicles; pitch, roll, yaw and thrust (Figure 8.19). Although normally operated manually by remote control, this approach can be programmed with each motion method weighted proportionally to create a diverse range of motion behaviours.

Similar to *Robotic Prometheus*, S-S_DRL research explored the programming of bespoke motion behaviours to support spinning, twisting and other actions with variable agility during a manufacturing task, and parallel investigations into how such movements might strategise manufacturing concerns such as additive manufacturing, to produce intricate designs from minimal tool-path motions. For example, a wax cylinder was fabricated by a simple helical movement, where material interconnects between circular passes due to the temporal time frame the wax takes to phase-change to a solid while being subjected to gravity. Extending to more autonomous behavioural approaches, similar motion behaviours can respond in real-time to computer vision feedback to achieve line-following capabilities (Figure 8.20-23) that support stigmergic approaches to building, similar to methods employed by social insects, where a robot is attracted to follow the path of

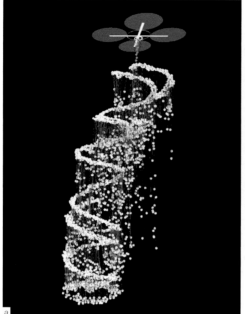

▲

8.24 Manufacturing Behaviour
a) Simulation of quadcopter depositing molten wax, b) molten wax prototype created using a simple helical trajectory.conditions.

previously built material (employed in *Plinθos: Aerial Sand Printing* project, discussed in Chapter Nine (Figure 9.32)). Quadcopters were also tasked with recognising objects or markers that they could move towards or around in programmed motion sequences that adapt to perceived conditions in real-time (Figures 8.20-8.23). These behaviours are used in *The Thread* (Figure 9.36) and *Aerial Floss* (Figure 9.40) projects described in Chapter Nine.

The relatively limited research undertaken into the use of aerial robots for construction activities (Zhang et al. 2022; Braithwaite et al. 2018; Augugliaro et al. 2013; Wood et al. 2019) has primarily relied on external motion capture systems to inform a quadcopter of its spatial location in Cartesian coordinates and typically employs a low-level controller (such as a PID or MPC)[1] to control the robot moving through a series of pre-determined waypoint positions. While such approaches could program a robot to spin, the use of spinning and other agile behaviours in S-S_DRL-programmed robots operated in a fundamentally different manner (Stuart-Smith 2016). Not reliant on external motion capture systems, S-S_DRL robots leveraged more degrees of autonomy and alternative means of environmental awareness. Throughout S-S_DRL research demonstrations, quadcopters ran a custom-developed ROS™ agent-based program to determine their motion behaviour and relied on computer vision, an intertial measurement unit (IMU) and other sensor-data fusion to estimate their spatial location and orientation using a form of Simultaneous Localisation and Mapping (SLAM) (Bowman et al. 2017). Although these experiments utilised an off-board computer connected to the robot over WiFi to program semi-autonomous activities, they could equally have operated using only an onboard computer, allowing the robots to be considered fully autonomous.

S-S_DRL research projects leveraged behaviour-based actions that operated through bottom-up event-based rules that enabled designs to arise from several robots' behaviours over time. Undertaken within a bare-bones research and teaching operation at the AA, these examples operated as proofs of concept, extrapolated within computer simulations and speculative design activities. The approach explored embodied, situated robot behaviour, to support embodied, situated design possibilities, where local adaptive behaviours enabled site-specific design responses. Although not necessarily an advantageous augmentation for all building applications, this semi-autonomous design approach might offer more sensitivity, adaptation and engagement potentials compared to alternative forms of computational design, particularly in temporal scenarios where robot-sensed spatial data might provide greater relevance than data collected over longer time-frames. Some of S-S_DRL's temporal, speculative designs are presented in the next chapter alongside funded research into aerial additive manufacturing that showcases the world's first demonstration of untethered additive manufacturing in flight.

In this chapter's explorations into the fabrication of individual parts, the off-site assembly of building modules, and on-site aerial construction, behaviour-based programming enabled creative engagement with manufacturing operations. With algorithms controlling robot/s' actions, these programs were embodied, while the use of sensor feedback from materials or the environment rendered these processes as also situated. This situated and embodied approach supported an adaptive manufacturing capability that was demonstrated to produce distinctive design character, and to support greater levels of productivity, offering an industrially scalable approach to the manufacturing of bespoke designs. This work offers possibilities for addressing several of the societal, environmental and economic challenges raised in Chapter Two, and in realising highly specific design responses discussed in the previous chapter. The next chapter extends these capabilities into live acts of construction where embodied robot systems are employed to adaptively build, and in some cases, to design through a situated act of construction.

▼
8.25 Manufacturing Behaviour
A quadcopter autonomously moving, both turning and yawing in a series of tight circular motions while maintaining altitude.

Adaptive Manufacturing

Notes

1. Such as a Proportional-Integral-Derivative (PID) or Model-Predictive-Controller (MPC).

2. Localisation is a robotics term to describe ascertaining the location (position) and orientation of a robot in reference to its initial position and a current position, typically defined by Cartesian or geographic coordinates.

3. Masters of Science in Design: Robotics and Autonomous Systems program (MSD-RAS) is a post-professional degree program in the Weitzman School of Design, University of Pennsylvania.

4. MSD-RAS course ARCH802: Material Agencies Robotics & Design Lab. Instructor: Robert Stuart-Smith. TA: Patrick Danahy. 2021.

5. SituatedFabrications Fabrication Studio. Visiting Professor: Robert Stuart-Smith. Academic Chair: Marjan Colletti. Academic Staff: Johannes Ladinig, Georg Grasser, Pedja Gavrilovic

6. Chapter Two outlined needs and opportunities for greater automation in the construction industry to address current rates of low productivity, skilled labour shortages, hazardous working conditions, excessive material waste, high-cost, unpredictable time-frames, and environmental impact, and other factors.

7. Construction company Laing O'Rourke successfully built off-site 85% of London's 47-storey, £340 million Leadenhall Building, but lost over £53 million in 2015 due to challenges faced in their pioneering work on three such projects (Sweet 2015). While Laing O'Rourke is optimistic for the future, in the US, Katterra cannot be. The vertically integrated design and off-site manufacturing company raised $1.6 billion on a similar vision but went bankrupt in 2021 due to what has been attributed to a lack of focus, scaling too fast without optimising their manufacturing operations (buildd, 2022).

Project Credits

Robotic Prometheus (2021): MSD-RAS program thesis project, University of Pennsylvania, Weitzman School of Design. Students: Grey Wartinger, Matthew White, Jiansong Yuan. Instructors: Robert Stuart-Smith, Nathan King, Billie Faircloth, Jose-Luis García del Castillo y López, Jeffrey Anderson. TA: Patrick Danahy.

SituatedFabrications Fabrication Studio (2016): Visiting Professor: Robert Stuart-Smith. Academic Chair: Marjan Colletti. Academic Staff: Johannes Ladinig, Georg Grasser, Pedja Gavrilovic. Students: Andreas Auer, Monique banks, Marcus Bernhard, Thomas Bortondello, Marc Differding, Christophe Fanck, Konstantin Jauck, Emanuel Kravanja, Jil Medinger, Sonia Molina-Gil, Pol Olk, Michael Schwaiger, Mario Shaaya, Melina Stefanova, David Stieler, Theresa Uitz, Matthias Vinatzer, Alexandra Zeinhofer.

Visceral/L (2022): Autonomous Manufacturing Lab, University of Pennsylvania (AML-Penn). Research Lead: Robert Stuart-Smith, Research Assistants: Riley Studebaker, Mingyang Yuan, Nicholas Houser, Jeffrey Liao.

MAP (2018): Autonomous Manufacturing Lab, University College London, Department of Computer Science (AML-UCL). Research Leaders: Vijay Pawar and Robert Stuart-Smith. Researchers: Julius Sustarevas, Daniel Butters, Mohammad Hammid, George Dwyer.

Youwasps (2019): Autonomous Manufacturing Lab, University College London, Department of Computer Science (AML-UCL). Research Leaders: Vijay Pawar and Robert Stuart-Smith. Collaborator: David Gerber. Researchers: Julius Sustarevas, Benjamin K. X. Tan.

M-RAM (2022): Autonomous Manufacturing Lab, University College London, Department of Computer Science (AML-UCL). Research Leaders: Robert Stuart-Smith, Vijay Pawar, Mirko Kovac, Jacqueline Glass. Researchers: Alec Burns, Thomas Legleu, Harvey Stedman, Ziwen Lu.

Husky+ (2021): Autonomous Manufacturing Lab, University College London, Department of Computer Science (AML-UCL). Research Leaders: Vijay Pawar and Robert Stuart-Smith. Researchers: Harvey Stedman, Ziwen Lu, Alec Burns, Daniel Butters.

Image Credits

Figures 8.1, 8.8–8.14: Robert Stuart-Smith, Autonomous Manufacturing Lab, University of Pennsylvania.

Figures 8.2–8.3: Robotic Prometheus: MSD-RAS program thesis project, University of Pennsylvania, Weitzman School of Design. Students: Grey Wartinger, Matthew White, Jiansong Yuan. Instructors: Robert Stuart-Smith, Nathan King, Billie Faircloth, Jose-Luis García del Castillo y López, Jeffrey Anderson. TA: Patrick Danahy.

Figures 8.4, 8.6-8.7: SituatedFabrications Fabrication Studio. Visiting Professor: Robert Stuart-Smith. Academic Chair: Marjan Colletti. Academic Staff: Johannes Ladinig, Georg Grasser, Pedja Gavrilovic. Students: Andreas Auer, Monique banks, Marcus Bernhard, Thomas Bortondello, Marc Differding, Christophe Fanck, Konstantin Jauck, Emanuel Kravanja, Jil Medinger, Sonia Molina-Gil, Pol Olk, Michael Schwaiger, Mario Shaaya, Melina Stefanova, David Stieler, Theresa Uitz, Matthias Vinatzer, Alexandra Zeinhofer.

Figures 8.5, 8.19: Robert Stuart-Smith.

Figures 8.15–8.18: Autonomous Manufacturing Lab, University College London, Robert Stuart-Smith and Vijay Pawar.

Figures 8.20-22: Aerial Floss. AA.DRL Studio Stuart-Smith, Students: Kai-Jui Tsao, Patchara Ruentongdee, Yuan Liu, Qiao Zhang. Supervisor: Robert Stuart-Smith. TAs: Tyson Hosmer, Manos Matsis.

Figure 8.23: Aerial Symbiosis. AA.DRL Studio Stuart-Smith, Students: Yuan Feng,Wenqian Huang, Jingjing Shao, Sujitha Sundraraj. Supervisor: Robert Stuart-Smith. TAs: Tyson Hosmer, Melhem Sfeir.

Figures 8.24-8.25: Minusplus. AA.DRL Studio Stuart-Smith, Students: Alejandro Garcia Gadea, Ashwin Balaji Anandkumar, Chiara Leonzio, Martina Rosati. Supervisor: Robert Stuart-Smith. TAs: Tyson Hosmer, Melhem Sfeir.

References

Arkin, Ronald C. 1998. Behavior-Based Robotics. Bradford Book. MIT Press.

Augugliaro, Federico, Ammar Mirjan, Fabio Gramazio, Matthias Kohler, and Raffaello D'Andrea. 2013. "Building Tensile Structures with Flying Machines." In Intelligent Robots and Systems (IROS), 2013 IEEE/RSJ International Conference On, 3487–92.

Bowman, Sean L, Nikolay Atanasov, Kostas Daniilidis, and George J Pappas. 2017. "Probabilistic Data Association for Semantic SLAM." 2017 IEEE International Conference on Robotics and Automation (ICRA), 1722–29. http://www.cis.upenn.edu/~kostas/mypub.dir/bowman17icra.pdf.

Braithwaite, Adam, Talib Alhinai, Maximilian Haas-Heger, Edward McFarlane, and Mirko Kovač. 2018. "Tensile Web Construction and Perching with Nano Aerial Vehicles." In Robotics Research, 71–88. Springer.

Brayer, Marie-Ange. 2015. "Frank Gehry: The Interlacing of the Material and the Digital." In Frank Gehry, edited by A Lemonier and F Migayrou. Prestel.

Brooks, Rodney. 2002. Flesh and Machines. How Robots Will Change Us. Nelson.

Brooks, Rodney (MIT). 1985. "A Robust Layered Control System for a Mobile Robot. AI Memo 864." Ieeexplore.Ieee.Org, no. September.

buildd. 2022. "What Happened to Katerra - Katerra Problems?" Buildd. https://buildd.co/startup/failure-stories/katerra-problems.

Butters, Daniel, Emil T. Jonasson, Robert Stuart-Smith, and Vijay M. Pawar. 2019. "Efficient Environment Guided Approach for Exploration of Complex Environments." In IEEE International Conference on Intelligent Robots and Systems. doi:10.1109/IROS40897.2019.8968563.

Castillo y López, Jose Luis Garciá del. 2019. "Robot Ex Machina a Framework for Real-Time Robot Programming and Control." In Ubiquity and Autonomy - Paper Proceedings of the 39th Annual Conference of the Association for Computer Aided Design in Architecture, ACADIA 2019.

Clearpath Robotics Inc. 2023. "Husky UGV - Outdoor Field Research Robot by Clearpath." Accessed June 29. https://clearpathrobotics.com/husky-unmanned-ground-vehicle-robot/.

Editors of Encyclopaedia Britannica. 2021. "Proprioception." Encyclopaedia Britannica. December 29. https://www.britannica.com/science/proprioception.

Editors of Encyclopaedia Britannica. 2022. "Personal Computer." September 20. https://www.britannica.com/technology/personal-computer.

EPSRC EP/S031464/1. 2018. "EPSRC GoW: Applied Off-Site and On-Site Collective Multi-Robot Autonomous Building Manufacturing." https://gow.epsrc.ukri.org/NGBOViewGrant.aspx?GrantRef=EP/S031464/1.

Gordon, Brian. 2023. "Durham Startup BotBuilt Robots Can Help with Housing Shortage." News & Observer. July 2. https://www.newsobserver.com/news/business/real-estate-news/article276860473.html.

Harrison, H. 2014. Jackson Pollock: Phaidon Focus. Phaidon Press.

Holland, Brynn. 2018. "Human Computers: The Women of NASA." History Channel. August 22. https://www.history.com/news/human-computers-women-at-nasa.

Landau, E G. 2000. Jackson Pollock. Harry N. Abrams.

Lorenz, Edward N., and Robert C. Hilborn. 1995. "The Essence of Chaos." American Journal of Physics 63 (9). doi:10.1119/1.17820.

Ma, Zhao, Simon Duenser, Christian Schumacher, Romana Rust, Moritz Bächer, Fabio Gramazio, Matthias Kohler, and Stelian Coros. 2020. "RobotSculptor: Artist-Directed Robotic Sculpting of Clay." In Proceedings - SCF 2020: ACM Symposium on Computational Fabrication. Association for Computing Machinery, Inc. doi:10.1145/3424630.3425415.

Malafouris, Lambros. 2008. "At the Potter's Wheel: An Argument for Material Agency." In Material Agency, 19–36. Springer US. doi:10.1007/978-0-387-74711-8_2.

Matarić, Maja J. 2007. The Robotics Primer. Intelligent Robotics and Autonomous Agents. Cambridge University Press.

Michaux, Henri. 2000. Untitled Passages by Henri Michaux. Edited by M. Catherine de Zegher. Ann Arbour: Merrell.

Moura, Leonel. 2013. Robot Art: A New Kind of Art. Createspace Independent Pub.

Science and Technology Select Committee. 2018. "Off-Site Manufacture for Construction: Building for Change." London.

Stuart-Smith, Robert. 2016. "Behavioural Production: Autonomous Swarm-Constructed Architecture." Architectural Design 86 (2). John Wiley & Sons, Ltd: 54–59. doi:https://doi.org/10.1002/ad.2024.

———. 2018. "Approaching Natural Complexity: The Algorithmic Embodiment of Production." In Meeting Nature Halfway, edited by Marjan Colletti and Peter Massin, 260–69. Innsbruck University Press.

Stuart-Smith, Robert, Riley Studebaker, Mingyang Yuan, Nicholas Houser, and Jeffrey Liao. 2022. "Viscera/L: Speculations on an Embodied, Additive and Subtractive Manufactured Architecture." In Traits of Postdigital Neobaroque: Pre-Proceedings (PDNB), edited by Marjan Colletti and Laura Winterberg. Innsbruck: Universitat Innsbruck.

Sweet, Rod. 2015. "Laing O'Rourke Loses Millions on Modern Construction Techniques, but Won't Give Up." Global Construction Review. September 7. https://www.globalconstructionreview.com/laing-orourke-loses-millions-m7od7e7rn/.

Tan, Rachel, and Stylianos Dritsas. 2016. "Clay Robotics: Tool Making and Sculpting of Clay with a Six-Axis Robot." In CAADRIA 2016, 21st International Conference on Computer-Aided Architectural Design Research in Asia - Living Systems and Micro-Utopias: Towards Continuous Designing.

Walter, William Grey. 1953. The Living Brain. Norton.

Werner, Frank, and CoopHimmelblau. 2000. Covering + Exposing: The Architecture of Coop Himmelb(l)Au. Princeton Architectural Press.

Wiener, Norbert. 2019. Cybernetics or Control and Communication in the Animal and the Machine. doi:10.7551/mitpress/11810.001.0001.

Wiener, Norbert, and John von Neumann. 1949. "Cybernetics or Control and Communication in the Animal and the Machine." Physics Today 2 (5). AIP Publishing: 33–34. doi:10.1063/1.3066516.

Willmann, Jan, Michael Knauss, Tobias Bonwetsch, Anna Aleksandra Apolinarska, Fabio Gramazio, and Matthias Kohler. 2016. "Robotic Timber Construction - Expanding Additive Fabrication to New Dimensions." Automation in Construction. doi:10.1016/j.autcon.2015.09.011.

Wölfflin, H, P Murray, and K Simon. 1966. Renaissance and Baroque. Cornell University Press.

Wood, Dylan, Maria Yablonina, Miguel Aflalo, Jingcheng Chen, Behrooz Tahanzadeh, and Achim Menges. 2019. "Cyber Physical Macro Material as a UAV [Re]Configurable Architectural System." In Robotic Fabrication in Architecture, Art and Design 2018, 320–35. Springer International Publishing. doi:10.1007/978-3-319-92294-2_25.

Zhang, Ketao, Pisak Chermprayong, Feng Xiao, Dimos Tzoumanikas, Barrie Dams, Sebastian Kay, Basaran Bahadir Kocer, Alec Burns, Lachlan Orr, Talib Alhinai, Christopher Choi, Durgesh Dattatray Darekar, Wenbin Li, Steven Hirschmann, Valentina Soana, Shamsiah Awang Ngah, Clément Grillot, Sina Sareh, Ashutosh Choubey, Laura Margheri, Vijay M Pawar, Richard J Ball, Chris Williams, Paul Shepherd, Stefan Leutenegger, Robert Stuart-Smith & Mirko Kovac. 2022. "Aerial Additive Manufacturing with Multiple Autonomous Robots." Nature 609 (7928): 709–17. doi:10.1038/s41586-022-04988-4.

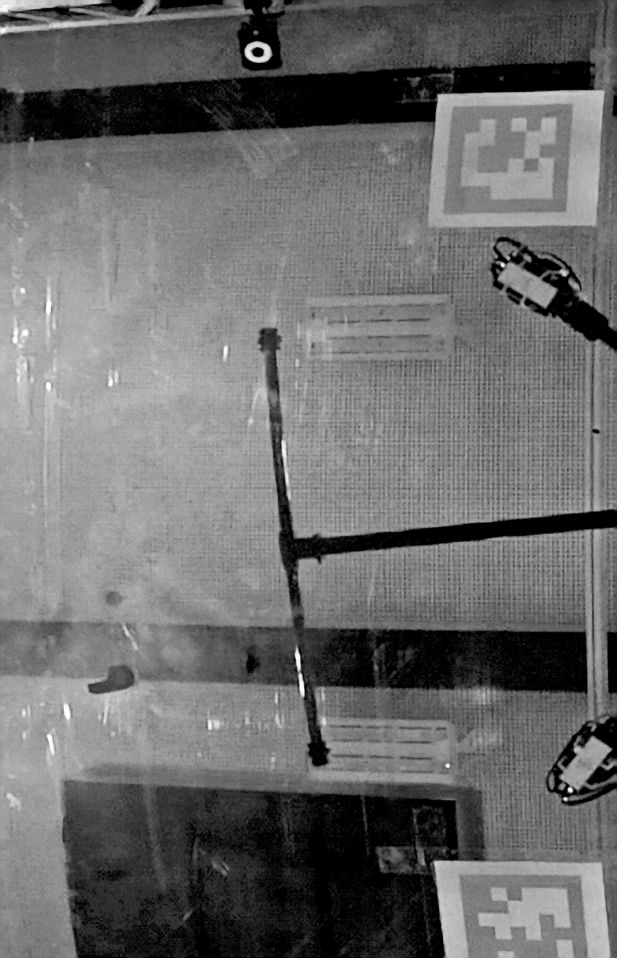

◀ ◀
9.1 Aerial AM
Photograph looking up through an acrylic table to a Buildrone additively manufacturing in flight.

Aerial Additive Manufacturing:
Explorations into Collective Robotic Construction

In warmer climates, there are cities comprised of megastructure towers, each housing millions, and boasting advanced environmental engineering solutions that rival that of any achieved in the history of architecture. Constructed from locally sourced, environmentally sustainable materials, these gigantic buildings require little energy to build and are continuously modified over time to adapt to the evolving needs of the occupants and changes in environmental conditions. Their mystifying complexity is even more interesting given that their design and construction is not planned in advance, or communicated in any drawings or 3D models. Instead, each building is realised through the collective act of construction by the inhabitants, none of whom comprehend the design of the overall structure. Termites build homes around one thousand times larger than their own size (Petersen et al. 2019), without a single crane, construction drawing or 3D BIM[1] file, and have been doing so for at least 135 million years (Krishna and Grimaldi 2003). Social insects including some termite and wasp species, utilise a building principle known as 'stigmergy' to indirectly build in reaction to each other and already built material through the sensing and deposition of pheromones as they build (Theraulaz and Bonabeau 1999; Grassé 1959). Computational models of stigmergic building have validated field and laboratory observations and demonstrated that such bottom-up event-based interactions can robustly give rise to complex, site-adaptive built habitats (Theraulaz and Bonabeau 1995; Bonabeau et al. 1998). Despite having considerably larger brains and heavy machinery at our disposal, along with the means to design, engineer and communicate plans in advance, our species struggles to build things efficiently at scales substantially greater in size than ourselves. The buildings we construct also do not achieve comparable levels of efficiency, complexity or variability.

Chapter Two described how current levels of construction productivity are inadequate to support present or future global needs. At a systemic level, present building methods do not scale activities effectively as buildings are scaled up, with larger projects suffering from longer delays (Farmer 2016, 16). More automation would help address these issues, however, due to the immense scale of buildings, automating on-site construction is extremely challenging. Recent developments in construction automation including Sky Factories (Bock and Linner 2016, 1) and on-site additive manufacturing (Dreith 2022) reduce construction time; however, their use of large gantry systems limits a building's size and shape to fit within the machinery's build envelope. Such large,

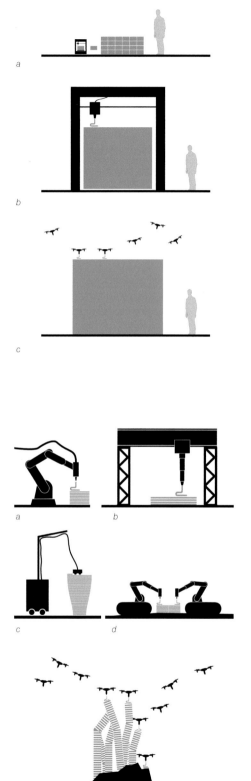

power-hungry equipment must also be transported to, and fit within a building site, and remain there for the duration of construction. Their sequential operation also constrains build times and scheduling. In contrast, social insects such as termites do a far better job of unbounded building, in parallel. For these reasons, researchers have been working on similar approaches using teams of mobile robots for quite some time.

Collective Robotic Construction (CRC)

Using a team of robots to build together may sound like science fiction yet substantial research has been done in this area for almost a half-century. MIT researchers Rodney Brooks and Maja Matarić investigated the possibility of using large teams of inexpensive, basic robots instead of expensive individual robots in the late 1980s (Brooks et al. 1990). As part of this research, Matarić's *'Nerd Herd'* robots demonstrated a simple building task (shovelling) by shuffling small pucks on a floor space (Matarić 1989). Recently, more elaborate demonstrations include: articulated robot arms collaborating on assembly tasks (Hosmer et al. 2020; Parascho et al. 2020), wall-climbing filament winding robots (Yablonina et al. 2017), small mobile brick-stacking and climbing robots (Petersen, Nagpal, and Werfel 2011), as well as aerial robots assembling elements (Lindsey, Mellinger, and Kumar 2011) or weaving structures (Braithwaite et al. 2018; Mirjan et al. 2016; Stuart-Smith 2016). This body of work is so extensive that a multidisciplinary group of collaborators including Drs. Kirsten Petersen, Nils Napp, Daniela Rus, Mirko Kovac and the author, teamed up to write a review paper on the subject in *Science Robotics*. We argued that Collective Robotic Construction (CRC) is a field of research in its own right, with its own goals and challenges (Petersen et al. 2019). Recognising CRC might not be appropriate for all building activities, we suggested CRC offers benefits for building in remote or hostile environments where human access is challenging, or in unknown and dynamically changing environments, where automated systems would not be sufficiently robust to work effectively. Examples included post-disaster reconstruction, nuclear disaster sites and building further afield – on the Moon or Mars (Petersen et al. 2019). There are no readily available CRC systems in operation within industry to date. As the

9.2 Large-Scale AM
To additively manufacture a large-scale object one must either; a) manufacture in many smaller parts, b) use an AM machine larger than the object, c) manufacture in parallel.

review paper highlighted, achieving application-ready CRC systems will require further development in diverse research strands in parallel. The work discussed in this chapter is no exception, involving large multi-disciplinary collaborations that demonstrate a proof-of-concept, and leave much work still to be done.

One key area of CRC research is in high-level planning, aimed at coordinating the actions of several robots working adjacent to one another concurrently. Most of the research described in the chapter is focused on high-level path-planning, building logistics and integrated design tasks, distinguishing the contribution outlined in this chapter from that of several collaborators. While CRC does not always include creative design activities, approaches to high-level mission planning and architectural design warrant integration and co-development as they inevitably impact each other. Also, by addressing these activities together, more creative freedoms are obtainable. This chapter explores several different relationships between these two activities, presenting work developed across both architectural and computer science research in the Autonomous Manufacturing Lab (AML) in University College London's Department of Computer Science (AML-UCL)[2] and the University of Pennsylvania's Department of Architecture (AML-Penn)[3].

Prior to research cited in this chapter, CRC had primarily focused on the assembly of discrete volumetric elements (Petersen, Nagpal, and Werfel 2011; Lindsey, Mellinger, Kumar 2011) and filament winding/weaving (Augugliaro et al. 2013; Stuart-Smith 2016), with little research undertaken into CRC mission planning for additive manufacturing (AM) (Kayser et al. 2018) or its relation to building designs. This chapter covers research that explores both activities.

Multi-Robot Additive Manufacturing

9.3 AM Technologies
Fixed build volume systems: a) Industrial Robot, b) gantry. Path-constrained: c) climbing robot AM, d) mobile ground AM Unbounded in 3-axis: e) Aerial AM

In Chapter Six, building-scale Material Extrusion (MEX) additive manufacturing (AM) was discussed, such as the continuous horizontal extrusion of concrete into vertically stacked contoured layers (Gibson et al. 2021). On-site MEX-AM methods typically use a gantry or crane to move an extrusion end-effector to any areas within the AM build envelope. However, as discussed in a recent paper in *Nature*, such AM systems' build envelopes are as limiting as that of a desktop 3D printer, only able to produce objects smaller than the machine (Figure 9.2b). Since construction sites vary in size, shape, and accessibility, it is challenging for such large equipment, and their installation, transport and power requirements to be practical for every use case, especially in remote locations (Zhang et al. 2022).

In *Nature*, we noted that while MEX-AM methods are faster than traditional construction approaches, a more flexible solution can be achieved by distributing the building task among several smaller mobile robots (Figure 9.2c). Although their reduced size may lead to a slower manufacturing rate, this can be offset by conducting parallel manufacturing activities (potentially powered by solar energy) (Zhang et al. 2022). Also, human workers cannot work safely in parallel to gantry-based robot systems as they must enter inside their manufacturing envelope. In contrast, a distributed robot system can potentially share a building site with human workers, dynamically adapting to avoid occupying the same space concurrently. Several mobile robot platforms have been developed to explore such

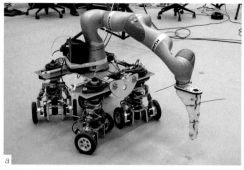
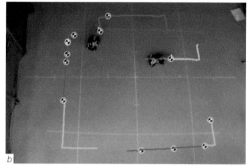

▲ ▲

9.4 Multi-Robot AM
Two AML-UCL custom-developed omni-directional robot AM platforms a) MAP#1 custom vehicle with Kuka iiwa arm, b) two Youwasps (modified Kuka Youbots) undertaking virtual AM.

▲

9.5 Aerial AM
Application Scenarios: a) post-disaster reconstruction in hard-to-access areas, b) distributed manufacture of larger buildings.

possibilities. However, most have operated with either tethered power or material supplies (Keating et al. 2017; Zhang et al. 2018), which limits their ability to operate over great distances or in large numbers without becoming tangled. Some are also constrained to move only on top of already-built material (Jokic et al. 2022; Kayser et al. 2018).

In the AML-UCL research lab, several untethered omnidirectional mobile ground robot systems were developed and tested for their ability to perform distributed, parallel additive manufacturing (Sustarevas et al. 2018; Sustarevas et al. 2019; Butters et al. 2019). AML's robot platforms couple together an industrial robot arm and an unmanned ground vehicle (UGV) robot with an integrated motion control strategy (Figure 9.4). In experiments these robots demonstrated an ability to collectively manufacture, autonomously navigating to and from material and energy supplies throughout a building process, and execute manufacturing tasks at horizontal scales larger than themselves, validating system scalability. These autonomous robots, however, can only operate where they have sufficient ground access. Limited to circulating around previously built material, multi-robot congestion would increase over the duration of building activities, while manufacturing height is also constrained by their vertical reach (Figure 9.3c-d). As an alternative approach, aerial robots can circulate more directly, passing over the congested space of a building site (Figure 9.3e). However, overcoming the immense challenges of additive manufacturing in flight required an ambitious collaborative effort.

Aerial Additive Manufacturing (Aerial AM)

The *Aerial AM* project emerged from a synthesis of the author's pilot research on a multi-agent aerial robot mission planner for design and construction (Stuart-Smith 2016) and Aerial Roboticist Prof. Mirko Kovac's research on aerial additive manufacturing spot-repair activities (Hunt et al. 2014). Prof. Kovac's research was conducted in the Aerial Robotics Lab at Imperial College London, and the author's through research and teaching in the Architectural Association's Design Research Laboratory: Studio Stuart-Smith[4] (that subsequently continued in AML-UCL and AML-Penn research labs). After running a series of joint workshops that established a shared conceptual approach to *Aerial AM*[5], a team of collaborators from diverse scientific fields[6] was assembled to co-lead the project including; Drs. Vijay M. Pawar, Stefan Leutenegger, Richard Ball, Chris Williams and Paul Shephard, together with our respective research teams at University College London, Imperial College London, University of Bath, University of Pennsylvania, Queen Mary University, Technical University of Munich, and Empa. Construction company Skanska, engineering firm Burohappold, 3D print company Ultimaker, and UK materials research institute BRE all operated as advisors on the project. The team was awarded UK Research and Innovation funding (EPSRC EP/N018494/1 2015) to develop the world's first working demonstration of untethered collective additive manufacturing in-flight. Demonstration achieved, the project was recently published as a cover article in *Nature* (Zhang et al. 2022).

Aerial AM was not proposed as a competitor to everyday construction methods, but to augment building capabilities for new applications and territories. Aerial robots can move around and directly over the top of obstacles that other robot platforms cannot, making their circulation more direct than other mobile robot platforms (Figure 9.3e). By incrementally printing in-flight with large teams of small aerial robots, new possibilities for building in remote or hard-to-access locations was envisaged, for applications such as post-disaster reconstruction, or building or repairing at height (Figure 9.5) (Zhang et al. 2022).

▶

9.6 *Aerial AM*
Aerial Robot Platforms: a) Buildrone, b) Scandrone, c) 2m high Aerial AM Foam Cylinder.

a

b

c

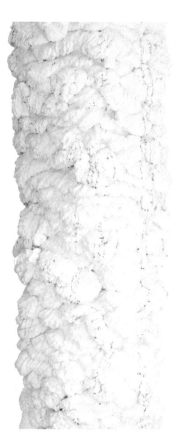

a   b

▲
9.7 Aerial AM
a) manufactured cylinder is several times larger than the drones, demonstrating scalability of the manufacturing method, b) close-up of cylinder print showing straight profile despite variation in individual layers.

To demonstrate *Aerial AM*, substantial challenges had to be overcome that are not faced in other forms of additive manufacturing. To state the obvious, it is difficult to deposit materially precisely while flying. Aerial robots are unstable in flight, tilt when moving, drift when trying to hold position, and are affected by wind or flight-induced air turbulence (downwash). They are typically flown individually by remote control, and no solutions to swarm control were available that supported the project's goals. At small sizes suited for indoor development and testing[7], aerial robots (or "drones") such as a quadcopter or hexacopter, also have limited battery power and payload capacity[8] to support material and extrusion equipment, making the lifting and extruding of materials challenging. To overcome these limitations, a custom aerial robot platform and controller needed to be developed, capable of flying and manufacturing with precision. To address the small payload constraint, multiple robots could build incrementally in parallel, requiring team-based coordination, collision avoidance and path planning. The payload limitations also suggested that lightweight high-volume materials were preferable, while most building-suitable materials were heavy. Material properties also impact extrusion hardware, so parallel development was required in both.

A custom-built *'Buildrone'* quadcopter was engineered to additively manufacture with an expanding polyurethane foam (Figure 9.6a). Beyond its established use in construction as insulation and permanent formwork for cast-in-place concrete, the lightweight foam also had a smaller transportation volume compared to its extruded volume, making

▲

9.8 Aerial AM
*a-c) time-lapse of
concrete Aerial AM, d)
manufactured outcome.*

it ideal for rapid proof-of-concept prints (Zhang et al. 2022). To ensure flights could accurately follow Cartesian coordinate waypoints, motion-capture cameras[9] tracked and informed the Buildrone of position and orientation. Despite this, commercial quadcopter controllers introduced position errors or noise, affecting accuracy. To overcome this, a model-predictive controller (MPC) was developed that enabled a Buildrone to undertake in-flight additive manufacturing with greater precision (Tzoumanikas, Yan, and Leutenegger 2020; Leutenegger et al. 2014).

A Buildrone was tasked with additive manufacturing in flight, a two-metre high, hollow cylinder. However, in early tests, the rapid expansion of the polyurethane foam produced an irregular deposition layer with an inconsistent build height. Repeatedly manufacturing on top of this variable outcome produced an accumulative deviation error that was likely to cause the cylinder to collapse after a certain number of layers. To address this, a scanning-in-the-loop method was developed that adjusted each manufacturing trajectory's waypoints and velocity to compensate for dimensional inconsistencies in a previous layer[10] (Zhang et al. 2022). A Buildrone additively manufactured in horizontal contour layers, while a second aerial robot *'Scandrone'* – scanned each layer and calculated adjustments for subsequent Buildrone manufacturing trajectories (Figure 9.6b). Deployed together, the two drones successfully manufactured a 2m high cylinder, taking alternating flights (Figure 9.6c) (Zhang et al. 2022). Despite substantial variation in each foam layer's height and position, the adaptive manufacturing

Aerial Additive Manufacturing

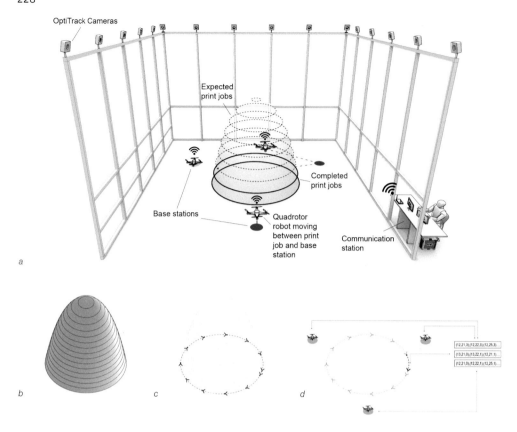

OptiTrack Cameras

Expected print jobs

Completed print jobs

Base stations

Quadrotor robot moving between print job and base station

Communication station

a

b

c

d

▲
9.9 Aerial AM
a) Flight-testing arena with Optitrack/Vicon motion capture system. Distributed manufacturing strategy: b) contoured geometry, c) contour divided into manufacturing lengths, d) each manufacturing job consists of a sequence of waypoints.

approach helped the drones adhere to the intended cylindrical geometry (Figure 9.7b). With error accumulation mitigated, the 2m high cylinder demonstrated that Aerial AM can support manufacturing at significantly larger scales than the Buildrone (Figure 9.7a). The tolerances involved, however, were insufficient for more filagree manufacturing applications.

To demonstrate suitability for various building applications, printing with a cementitious material is advantageous given its widespread use in construction. However, the weight, viscosity, curing time and aggregate size of available materials were impractical for aerial transport and extrusion with miniaturised hardware. A material specifically designed for *Aerial AM* had to be developed. It needed to have low viscosity to minimise hardware weight and energy requirements, while still having enough viscosity to maintain shape. The material also had to cure quickly after extrusion and meet structural requirements. A cementitious material was engineered to meet these challenging requirements, employing sheer thinning to reduce resistance during extrusion (Dams et al. 2017). Custom extrusion hardware was developed in parallel, and optimised for the new material. With an 8mm extrusion diameter, greater precision was also required. A lightweight delta-arm manipulator was developed to compensate for the robot's drift by dynamically correcting the extrusion nozzle position during flights (Chermprayong et al. 2019). These innovations enabled the additive manufacturing of a 300mm-high cylinder in-flight by two Buildrones undertaking alternating flights (Figure 9.8) (Zhang et al. 2022). The foam and concrete cylinders are demonstrations of the first successful untethered additive manufacturing in-flight. The coordination of their manufacturing was also achieved within a novel approach to high-level mission planning tailored to collective aerial additive manufacturing.

Collective Aerial AM Software Framework

A Collective *Aerial AM* software framework was developed to enable several Buildrones to autonomously manufacture in parallel (Figure 9.11) (Zhang et al. 2022). The framework functions similarly to desktop 3D printer software by building a geometry through the manufacturing of ascending contours. However, to support multiple robots manufacturing simultaneously, each contour was divided into several discrete jobs, sized to the maximum manufacturing length achievable by a Buildrone in one flight (Figure 9.9). As manufacturing in-flight is prone to uncertainties such as the robustness of robot platforms, adverse weather conditions, or other dynamic events that could disrupt a predetermined manufacturing sequence, adaptive solutions were necessary to provide redundancy to robots and building operations.

Unlike assembly-based aerial CRC demonstrations that have typically employed specified flight corridors (Augugliaro et al. 2013; Lindsey, Mellinger, and Kumar 2011), the framework allows for more flexible flight path planning and task determination. Mission planning is accomplished through multi-agent programming. Each Buildrone autonomously decides which manufacturing jobs to undertake and how to navigate to and from a job. Base stations are distributed at an offset distance from the build area where Buildrones can replenish material or power supplies (Zhang et al. 2022). Buildrones aim to take a direct route to their destination while dynamically avoiding each other and already-built work (Figures 9.9, 9.10). They will abort a job if there is too much congestion or are running low on power or material supply, leaving another robot to complete the remaining part of an aborted job. In the Collective *Aerial AM* framework, manufacturing is achieved through the collective behaviour of robot agents (Zhang et al. 2022).

▼
9.10 Aerial AM
Light-trace time-lapse photograph of several aerial robots manufacturing a dome-like geometry in parallel.

230

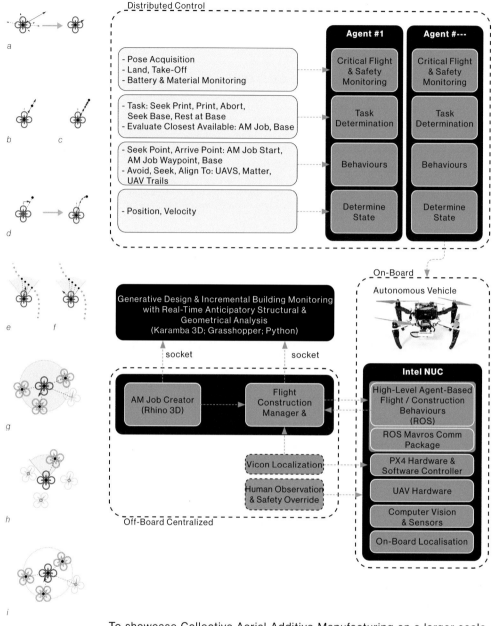

To showcase Collective Aerial Additive Manufacturing on a larger scale, an alternative method to concrete or foam manufacturing was pursued. A time-lapse video was created using a team of Scandrones (Figure 9.10). An LED light mounted on each ScanDrone shone with colours representing manufacturing and non-manufacturing activities as they built a virtual dome in flight (Zhang et al. 2022). Computer simulations using the software framework further demonstrated an adaptable, efficient, and scalable approach to swarm-based, distributed manufacturing, where system robustness was not dependent on the continued operation of any individual robot. Different population sizes of Buildrones were simulated manufacturing several scales of dome and cylinder geometries of comparable radius and manufacturing length. In these simulations, larger-sized geometries could accommodate more Buildrones operating without

◀

9.11 Aerial AM
*Collective Aerial AM
Software Framework.*

◀◀

9.12 Aerial AM
*Aerial robot agent-
based steering
behaviours.
a) sum vector influences
and limit velocity, b)
seek point vs, c) arrive
point, d) steering vector
calculation, seek (e)
and align (f) to
already 3D printed
matter and robots'
previous flight trails,
g) radius threshold for
neighbours, matter, h)
angle threshold field-of-
view, i) seek neighbours,
j) avoid neighbours, k)
align with neighbours.*

congestion, increasing system efficiency. The active robot population quickly adapted to the size of a cylinder and remained constant throughout its manufacturing. While for a dome, Buildrones gradually retired to avoid congestion as manufacturing approached completion (Figure 9.15). This decrease in population closely matched the reduction in footprint area in higher-level contours of the domes, showing a strong correlation between collective behaviour, geometry and system efficiency, irrespective of Buildrones having no global knowledge of the geometry they manufactured (Zhang et al. 2022). Simulations of more complex designs further demonstrate the versatility of Collective *Aerial AM* as a design-agnostic manufacturing approach (Figure 9.16) (Stuart-Smith et al. 2023).

The Collective *Aerial AM* software framework enables distributed, concurrent manufacturing of a geometry with a variable number of robots. Buildrones that are manufacturing have right-of-way while they follow a predetermined flight path. In contrast, Buildrones not manufacturing dynamically avoid collisions and plan their flight path in real-time using vector-based steering behaviours. This is achieved by continuously providing small incremental waypoint goals to the robot's low-level controller at a specific magnitude and frequency. Recently calibrated with a commercial controller[11] in lieu of the custom MPC, sufficient accuracy was achieved to enable this adaptive flight mode to potentially also be used during manufacturing if desired[12] (Figures 9.13, 9.14) (Stuart-Smith et al. 2023).

▶

9.13 Aerial AM
*Incremental adaptive
flight tests using
one scandrone: a)
trajectories from 53 flight
tests, b) high accuracy
trajectories, c) top view
of an accurate trajectory.*

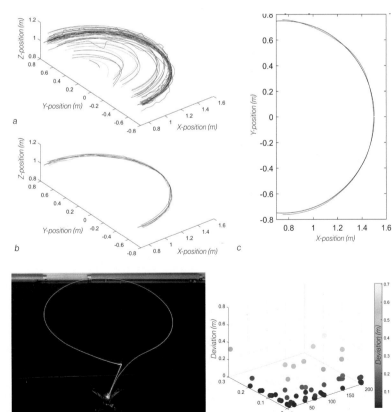

▶

9.14 Aerial AM
*a) Time-lapse photo of
one flight trajectory, b)
position error in relation
to waypoint frequency
and magnitude.*

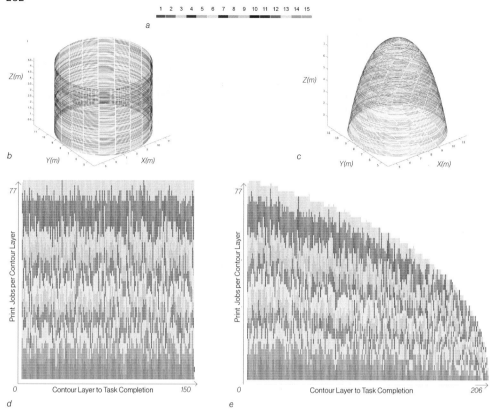

▲

9.15 Aerial AM
Simulated task distribution
across a population of 15
aerial robots. Comparison
between 12m diameter
cylinder and dome-like
geometries with equal
total print length:
a) each colour represents
an individual robot, b)
cylinder contours, c) dome
contours, d) graph view of
"b", d) graph view of "c".

Collective *Aerial AM* operates primarily through agent-based autonomous decisions similar to those discussed in earlier chapters. Each agent's behaviour adapts to locally perceived events. Although decisions are de-centralised, robots rely on centralised data for external localisation from a motion capture system (Figure 9.9). From the scientific research community's perspective, these robots are therefore considered semi-autonomous. Current research is extending capabilities for outdoor applications, employing more autonomous means of localisation. Collective *Aerial AM* allows a team of robots to autonomously build a predefined design, but not all designs are suitable for this approach. Consideration for *Aerial AM's* building constraints and opportunities needs further exploration. While research continues towards more practical design and build applications, this chapter also includes a series of speculative projects that conceptually leverage Collective *Aerial AM's* incremental, distributed approach to building and its impact on design possibilities.

Collective Aerial AM: Incremental Construction

In addition to developing a physically capable *Aerial AM* system, a series of speculative projects explored the creative potential of collective construction. One such project addressed challenges faced in the sequential act of building. Incremental building strategies for assembly-based construction have been demonstrated in CRC research (Andreen et al. 2016; Werfel, Petersen, and Nagpal 2011); however, no similar approaches have explored AM architecture or addressed the inclined and horizontal spans used in most buildings today. Many buildings require temporary support scaffolds during construction to avoid collapse, which adds complexity and increases transportation volume and costs. Increases in building surface area also produce significant increases in scaffolding volume, which is not practical. To avoid such concerns, building designs typically constrain on-site AM to vertically extruded geometries (Chen and Yossef 2015; Sevenson 2015). Unfortunately, this requires horizontally spanning elements like floors or roofs to be built by other methods, making full on-site automation difficult and remote applications of *Aerial AM* less feasible.

A building with vertical walls and horizontal spanning elements is also less structurally and materially efficient compared to three-dimensional curvilinear geometries like 3D shell structures, which provide spatial enclosure using only a thin layer of material. Funicular shell structures designed by Gaudi, Block and others (Cuito and Montes 2003; Rippmann and Block 2013) use curved surfaces to reduce horizontal thrust, enabling a breathtakingly thin and structurally efficient construction to be achieved. However, these geometries also require temporary support scaffolding during construction. To address this, a 3D shell design approach was developed that supports incremental manufacture. The approach allows the Collective *Aerial AM* construction of larger-span, material-efficient three-dimensional surface designs without the need for temporary support scaffolds. First, a custom funicular design method generates diverse 3D shells from simple two-dimensional surface patterns. Then, mid-construction collapse is prevented by locally reinforcing the designed surface with additional ribs or vertical supports (Figures 9.17, 9.20). The approach can be used to modify a design before or during construction.

▼

9.16 Aerial AM
Geometrically agnostic construction: simulated manufacturing of two different multi-genus topological geometries (a,b).

Building upon multi-agent design-simulation methods discussed in Chapter Six, an initial shell surface design is analysed to identify areas that require more vertical support or reinforcement[13]. A swarm of virtual agents is then seeded on the surface and moves towards identified new support locations, weighting several vector-based steering behaviours against multiple design criteria (see Chapter Six). The agents' trajectories create a series of curves that connect to each other and back to the shell and are interpolated

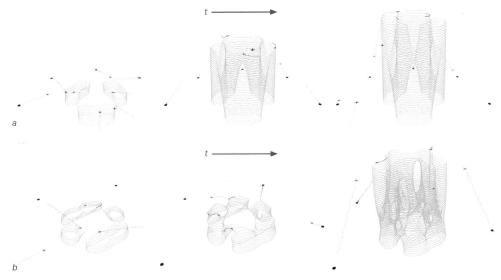

a

b

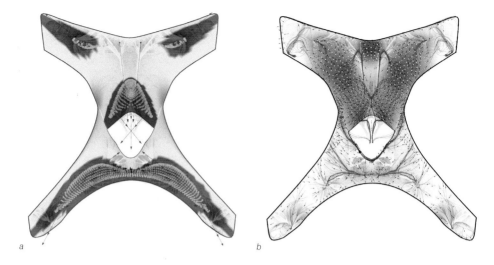

a

b

as a volumetric isosurface mesh together with the original surface, resulting in a ribbed shell with truss and column-like conditions. Structural analysis was performed on the design at intermediate construction states and the final built state. Results demonstrate significant improvements but also indicate the need for further material optimisation measures (Figure 9.19).

As material reinforcement was determined through local agent-based decisions, the process could be repeated several times throughout construction in response to real-time monitoring of building progress. A software pipeline was established between the Collective *Aerial AM* framework and the design and analysis method to support this in future research. Beyond the technical merits of the approach, the shell's aesthetic outcome is tightly related to its design and manufacturing approach (Figures 9.21, 9.22). Although behaviour-

▼
9.18 Incremental Construction
Top view of Aerial AM simulation (a-c).

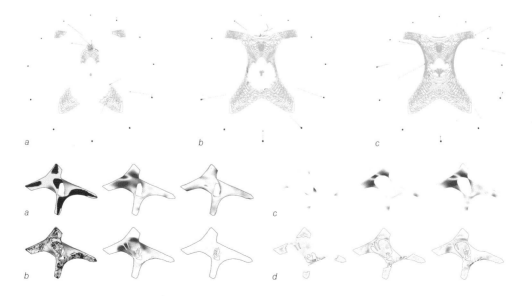

a

b

c

a

c

b

d

◀

9.17 Incremental Construction
Reflected ceiling plan view of multi-agent local reinforcement method: (a) additional support vectors defined in shallow inclination regions; (b) agents connect high deflection areas to new support locations.

based programming was used for both design and construction activities, each operated independently. There are, however, creative benefits in unifying these two activities, allowing a design to arise from, or embody a building strategy.

Collective Aerial AM: Generative Design by Self-Organisation

As Buildrones can extrude a cementitious material, the aerial additive manufacture of large compression-based structures is theoretically possible. Ander et al. have provided theoretical proof that a tower could be built to an unlimited height by adhering to a number of geometric constraints that resist buckling by increasing the cross-sectional area and wall thickness towards the bottom of a tapering tower's structure, enabling structural stress to be sufficiently low and constant[14] (Ander et al. n.d.). In relation to this, structural engineers Chris Williams and Paul Shephard factored in the material properties of concrete and confirmed it is suitable to use as the structure for towers substantially taller than any towers built to date. Using a generative structural algorithm developed by Chris Williams, a design proposal for a tower with similar floor plate dimensions to New York's original World Trade Centre towers was created, reaching a height of two kilometres (Figure 9.23). While the practical feasibility of such a tall project was not explored, it

▶
9.20 Incremental Construction
Reflected ceiling plan

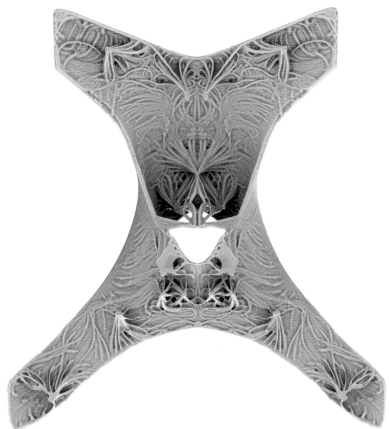

◀
9.19 Aerial AM Shell
Worm's-eye Comparative Analysis demonstrates reduction of collapse and deflection possibilities.

L to R: final inclination angles, structural deflection, principle stress, a) initial shell, b) locally reinforced shell.

Structural deflection at 3 incremental stages of construction L to R. Rows c) initial shell, d) locally reinforced shell.

Aerial Additive Manufacturing

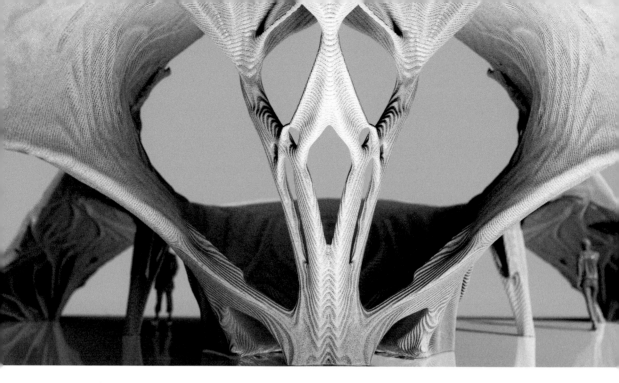

▲

9.21 Aerial AM Shell
*View of 3D shell
design for incremental
construction by Collective
Aerial AM. Additional ribs
and supports mitigate
for instability throughout
construction.*

is worth noting that elevators capable of 1km high continuous runs have already been developed (Staff Writer 2013), and tall buildings already use sky lobbies to connect multiple elevators.

As the tower geometry is relatively vertical, it could theoretically be additively manufactured on-site without scaffolding. However, due to the tower's inclined form and branching topology, it would be challenging to additively manufacture using incremental building approaches such as jump-form or slip-form construction techniques that are designed for vertical extrusion[15]. Their modular equipment could not be easily adapted for building multiple inclined branches simultaneously or adjusted to the variable and inclined geometry of each branch. Similarly, a crane-like AM system would be challenging to reposition or multiply across several branching and converging volumes over the height of the tower. To overcome these limitations, a gantry-based AM system would need a larger build volume than the two-kilometre-high tower itself, which is not feasible. On the other hand, a distributed, collective *Aerial AM* approach can scale to such heights without requiring additional infrastructure, demonstrating a synthesis between the structural and formal tower design and its means of autonomous multi-robot manufacture.

To allow for natural daylighting and views, the exterior surface of the tower could not be completely solid (Figure 9.24). A speculative design for a structural exoskeleton was developed that leveraged the logistics of aerial robots to operate not only as builders but also as design agents. Taking the *Aerial AM* research demonstrations as a departure point, Scandrones operated with design agency as they ascended ahead of Buildrone additive manufacturing activities. A simulation was developed of aerial robots undertaking the construction of the tower while their collective interactions also informed the design of the exoskeleton (Figures 9.25, 9.26). The construction simulation was, therefore, also a generative design algorithm. Like the Collective *Aerial AM* framework, it was envisaged that the aerial robots had access to a shared virtual

3D model of the tower. Initially distributed at equal distances around the virtual tower's base, a simulation of Scandrones ascended in flight, constrained to the tower surface geometry while moving in coordination with one another to define a networked exoskeleton path that could potentially support floor plates inside. Buildrone agents followed below, manufacturing a volumetric mesh interpolation of the aerial robots' trajectories in horizontal contours (Figure 9.25).

As the tower volume was developed using a structural algorithm, it embodied a relatively even distribution of stress forces across its exterior surface. Unlike a conventional tower, the design's increased surface area and volume near the base alleviated the need for an increase in the density of structural elements in lower regions to support the building's weight above. To ensure a relatively even distribution of the exoskeleton, local event-based agent rules informed Scandrone building behaviour. As Scandrones ascended the virtual building surface, they would move closer to each other if there were few neighbouring Scandrones nearby, and move apart if overcrowded. If the nearby population exceeded a certain threshold, a Scandrone would retire, while if the population fell below another threshold, an additional Scandrone would join (Figure 9.26). These programmed behaviours ensured that isolated vertical elements would branch while densely packed elements would converge, maintaining a relatively even density.

9.22 Aerial AM Shell
Internal branching column support at mid-span through the shell, with additional mid-span supports seen in the background.

Although the overall formal silhouette of the tower is constrained by the virtual structural design model, the exoskeleton is designed through the act of construction, by the actions of a swarm of *Aerial AM* Buildrones (Figures 9.25, 9.29). The resulting design is emergent, a complex yet coherently ordered outcome that was not predetermined or described in the agent's rules (Figure 9.30). It arises through the local interactions and self-organising behaviour of the robot agents over time. Such adaptive design-construction methods offer numerous possibilities, some of which were explored in a series of speculative design-research studio projects.

238

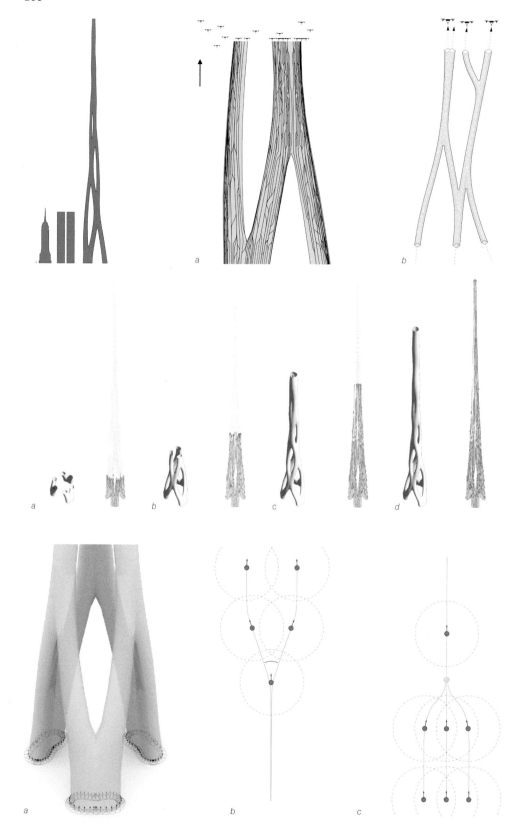

9.23 Aerial AM Tower
Scale comparison with
New York's Empire State
Building and World
Trade Center Towers.

9.24 Aerial AM Tower
Exoskeleton design
strategy: a) Scandrones
ascend to define
exoskeleton trajectories,
b) Buildrones manufacture
around trajectories in
horizontal contours.

9.25 Aerial AM Tower
(a-d) multi-agent
construction simulation
adapts to structural
surface geometry input.

9.26 Aerial AM Tower
Primary multi-agent
behaviours: a) initial
robot setout at tower
base, b) branching and c)
converging in relation to
robot population density.

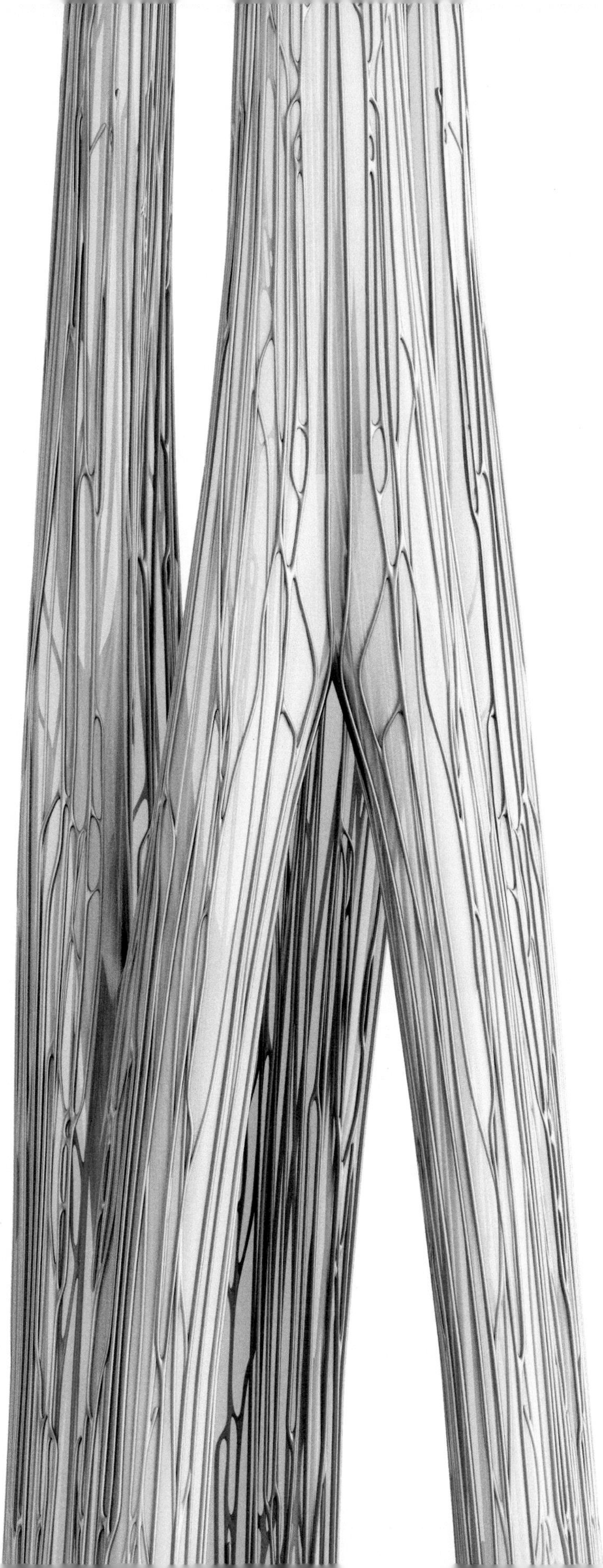

9.27 Aerial AM Tower
Partial perspective view of
tower exoskeleton design.

Simulating Aerial Swarm Construction

Early in the development of *Aerial AM*, Studio Stuart-Smith in the Architectural Association School of Architecture's Design Research Laboratory conducted research into aerial robot construction. Projects explored design through collective construction, engaging with diverse materials and environments in speculative application scenarios that were positioned to leverage *Aerial AM's* ability to operate in hard-to-access or remote locations such as arctic regions, caves, mountain villages, or deserts. Where possible, locally sourced materials were used, while design and construction operated through situated local events, suggesting solutions that theoretically enabled construction to be undertaken autonomously without supervision. Projects leveraged collective robot behaviour to create emergent design outcomes that arose from robot agents' spatial interactions (similar to those described earlier). Site, materials, and build process were integral to each design outcome. An *Aerial AM* Polar Research Center[16] leveraged swarm behaviour to create a cavity-wall shell from locally collected ice (Figure 9.32). Large-scale desert shelters[17] were conceived that used sand dunes as a temporary support scaffold and build material for a binder-jetting AM process. *Aerial AM* was directed towards the minimal task of depositing a water-soluble salt-solution binding agent that could transform part of a sand dune into a 3D printed filagree structure after it was excavated out of the dune (Figure 9.33). A third project sought to build circulation routes through the earth's most unchartered underground caves[18], employing collective robot spatial mapping to identify viable support surfaces, and an A* planning algorithm to path-plan an efficient circulation route for a networked series of elevated walkways. These informed a multi-agent design-construction method (Figure 9.31).

Two projects explored design solutions for inaccessible mountain regions. A mountain retreat project[19] involved a building strategy that adapted to terrain, wind and structural feedback. A generative surface growth algorithm operated in parallel to multi-agent construction operations, guiding early building to face oncoming winds to create a wind break for subsequent building (Figure 9.34). The second project speculated on how a pedestrian bridge could be constructed from two opposing sides as cantilevered

9.28 Aerial AM Tower
Perspective view of tower exoskeleton design

structures until both sides met to become a bridge[20], spanning like a beam (Figure 9.35). In a design-construction simulation, aerial robot agents commenced building in response to a relatively simple 3D design input. However, throughout building, progress is structurally analysed a few steps in advance. If a future collapse was deemed probable, the aerial robot agents could adapt their behaviour to build additional reinforcement supports to mitigate the future collapse before continuing. These actions shifted the design from its initial intention to a semi-emergent outcome, where improvised construction altered the project's aesthetic and rendered design and construction as inseparable from one another (Figure 9.36).

In these speculative projects, architecture students programmed multi-agent algorithms to operate as design-construction simulations, performed flight experiments with small aerial robots indoors, and developed material-specific manufacturing methods for proof-of-concept physical experiments. Although these activities were separate, they informed each other in a parallel development process. For instance, the *Aerial AM* Bridge project involved the development of a UV[21] light-cured resin additive manufacturing method for horizontal cantilevered conditions that was tested in several physical prototypes (Figure 9.35b-c), while the *Aerial AM* Dune Shelters project tested several material mixtures and end-effectors in physical flight tests. The *Aerial AM* Mountain Retreat project tested building behaviours in larger-scale cantilevered prototypes (Figure 9.34c). The development of material deposition hardware also influenced software and design iterations. Design speculations were grounded in technical discoveries and first-hand learned experience. Like the various approaches to material engagement discussed in Chapter Three, design operated partly through non-human agencies, where the designer's primary role was in the orchestration of events. This approach attains greater agency when autonomous sensing and action are employed. Two aerial weaving projects demonstrate this to different degrees.

▼
9.29 Aerial AM Tower
Worm's eye view of aerial robot multi-agent simulation. Colour represents changes in agent states that impact agent behaviour (a, b).

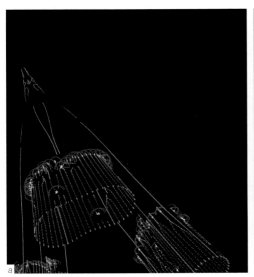
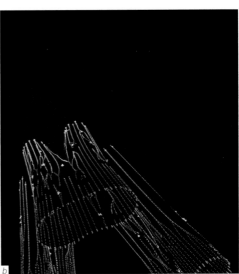

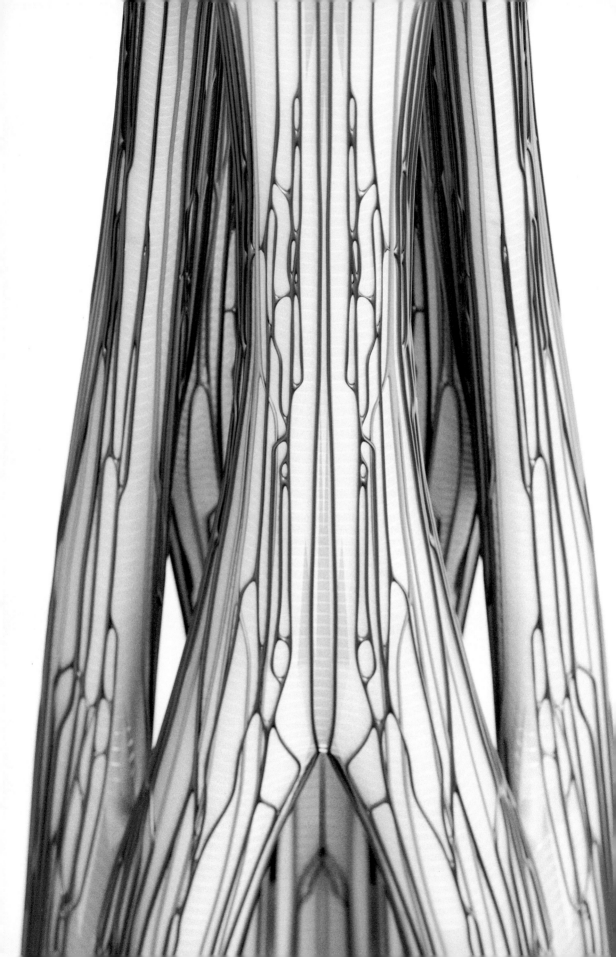

Exploring Autonomous Aerial Construction

The project *The Thread*[22] explored the use of aerial robots for constructing a tensile installation. Although similar methods were being researched simultaneously without our knowledge at Imperial College (Braithwaite et al. 2018) and ETZH (Augugliaro et al. 2013), the project was differentiated from these in three distinctive ways. Firstly, a generative design algorithm incorporating multi-agent and material-physics simulation guided aerial robot behaviour. Secondly, instead of relying on external motion capture equipment for 3D localisation, the aerial robots leveraged onboard sensing and Simultaneous Localisation and Mapping (PTAM SLAM) (Klein and Murray 2007). Lastly, due to limited resources, the project was carried out using low-budget quadcopters in a classroom setting.

Weaving a spatial structure in the air is quite challenging. Each aerial robot agent held a spool of thread whose end was fixed to a wall at a specified position. As the robots flew, they extended a length of thread out into space. To position an intersection of threads in 3D space, at least three lines of thread were needed to triangulate the position. To achieve this in a continuous flight sequence, aerial robot agents move together in formation, orbiting around each other to braid their threads together for a time, before branching away from each other to create a Y-shaped thread arrangement that could be stabilised by fixing each thread's endpoints, or otherwise be braided again with other threads to create more complex thread networks (Figure 9.38). As braiding formations were defined by event-based rules (rather than being explicitly defined by specific Cartesian coordinates), braiding behaviours could be deployed in different spatial settings. Simulations were developed and evaluated to test the robustness and design capabilities of braiding with varying quantities of agents starting from different locations (Figure 9.39).

The approach was then used to program a small team of aerial robots to construct a tensile installation indoors (the method would also work outdoors) (Figure 9.40). The resulting braided volume floated in between the room's walls, floors and ceiling, and held its position due to the triangulation of its braided threads. As a concept for tensile architecture, *The Thread* is scalable, with larger structures theoretically possible using the method. Little adaptation of the approach would be required to build a version of the project at scales suitable for human occupation[23] such as the inhabitable installations of artist Tomás Seraceno (Galansino 2020). *The Thread* was achieved by developing a design-construction sequence within a multi-agent simulation. A simulation outcome was then executed by multi-robot collective construction, leveraging autonomous robot sensing and localisation. A second project operates with greater degrees of autonomy, using computer vision to inform aerial robots' real-time design and construction decisions.

In *AerialFloss*[24], tethered helium balloons are used as a scaffold for a filament-winding aerial construction method. A lightweight and sticky polymer thread was created that could be used as a filament and wrapped around the tethers to create a temporary, lightweight open canopy to serve as an ethereal event space. Although the helium balloons were tethered to fixed locations on the ground, their 3D location could shift dramatically due to turbulence caused by the aerial robots, or wind if operating outdoors. Instead, each robot was programmed to recognise balloons and adjust their flight

244

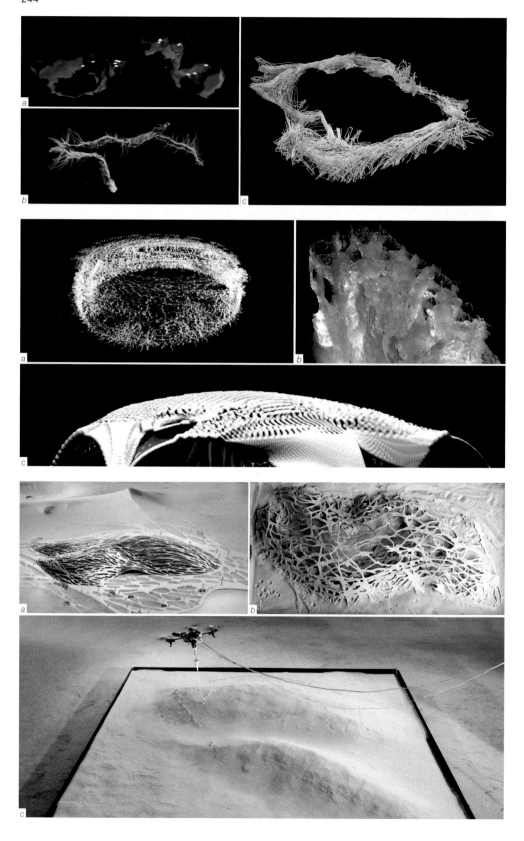

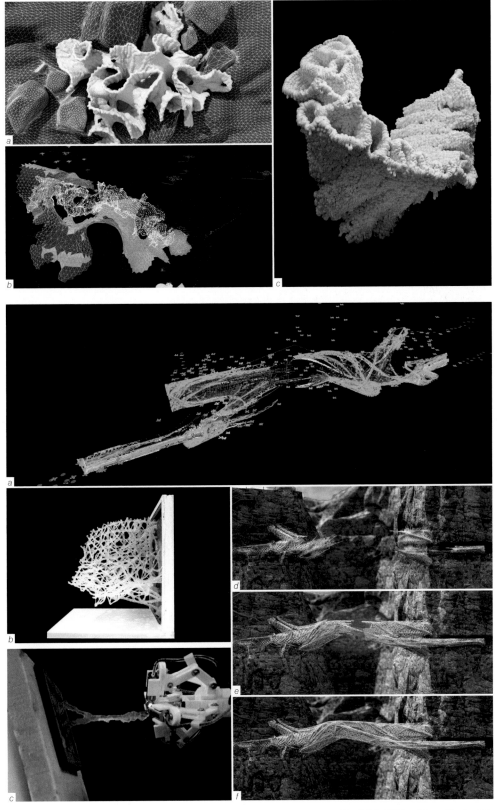

Aerial Additive Manufacturing

◄◄
9.34 Aerial AM Mountain Retreat
a) perspective view of retreat demonstrating formal adaptation to site, b) simulation of constructing into the wind, c) physical prototype

behaviour relative to perceived balloon locations, estimated using OpenCV (Bradski 2000) computer vision methods (Figure 9.41a). These behaviours supported aerial braiding and filament winding of the tethers to create a lightweight porous canopy. Computer vision was also utilised to provide the robots with autonomous material collection (Figure 9.41e) and thread collision avoidance capabilities (Figure 9.41b). The project was demonstrated through a small proof of concept indoors (Figure 9.42) and extrapolated on in digital simulations (Figures 9.43-9.46) and a speculative design proposal.

◄◄
9.35 Aerial AM Bridge
a) multi-agent simulation, b) cantilevered AM prototype, c) custom UV-cured resin extrusion tool, d-f) construction sequence

The *AerialFloss* design research began with physical experiments, testing event-based rules in robot manufacturing demonstrations (Figure 9.42). Knowledge gained was then encoded in a custom simulation software that included robot agent behaviour and a material-physics simulation of balloons and threads (Figure 9.43). A speculative design for the Burning Man Festival proposed a network of aerial canopies to be constructed during the festival in response to the varying size and location of gatherings (Figure 9.46). Like the other speculative projects, the design was created through a generative construction simulation (rather than being explicitly modelled). In *AerialFloss*, architecture was not only developed using event-based rules, it also functioned as an event-based system. *AerialFloss* builds on architectural precedents like Peter Cook's Instant City which was to be lifted in by airships (Cook and Archigram 1999, 96). In *AerialFloss*, however, autonomous robots adaptively construct an event-based architecture in response to human activities. To some extent, the spaces themselves could therefore be considered autonomous and adaptive, as they expand and contract relative to festival activities. The tethers were also envisioned to be dynamically repositioned and extended to great heights by autonomous ground robot vehicles, rendering the architecture almost unbounded (Figure 9.44).

Design Agency in Collective Robotic Systems

9.36 Aerial AM Bridge
Close-up partial view of the completed bridge.

Collective robotic construction (CRC) research is in its infancy while correlated architectural design research is especially nascent. The physically demonstrated *Aerial AM* system provides proof of concept for a CRC approach that supports unbounded, parallel construction at height, in hard-to-access areas, and potentially enables more site-adaptive building possibilities. A series of speculative projects explored some of these possibilities through physical flight demonstrations with a small number of robots, and computer simulations using larger populations.

In this chapter, collective behaviour supported design and construction activities that would be impossible to undertake with robots operating by remote (manual) or automated control. This work pushes beyond a human's capacity to engage with distributed, concurrent activities. Operating a large team of robots simultaneously by remote control (telemetry) is impractical due to limitations in human perception, cognition, and reaction times when controlling many moving objects. Assisted control approaches, such as a leader-follower framework (where a team of robots follows one remotely controlled robot), might provide some capacity to control a team of aerial robots, but such approaches do not support robots undertaking different tasks to build in parallel. Automated programs such as those typically employed on articulated industrial robot arms would also be insufficient for aerial robot operation. There is broad agreement in the scientific research community that mobile robots must be at least semi-autonomous to operate robustly in unknown or dynamically changing environments or where communication is unreliable or delayed.

9.37 The Thread
An aerial robot constructed tensile installation.

Degrees of autonomous programming are essential for controlling a large population of mobile robots and leveraging their collective behaviour. Therefore, CRC is inconceivable without some level of autonomous control. As was demonstrated in designs that integrated

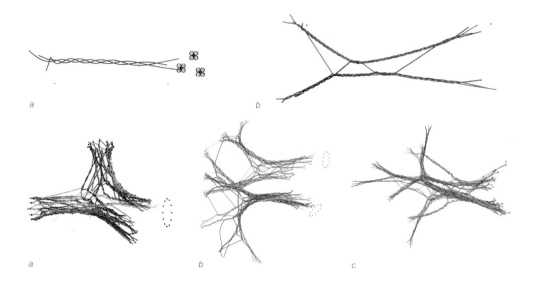

a

b

a

b

c

▲▲

9.38 The Thread
Multi-agent braiding
behaviours: a) braiding
in relation to adjacent
agents, b) two spanning
and interconnected
braids create a networked
tensile structure.

▲

9.39 The Thread
Different designs
emerge from the same
behaviours being applied
to different sites (a-c).

construction sequencing, structural stability, environmental and object dynamics, and other logistical considerations, a CRC architectural design approach also stands to benefit from incorporating some degree of autonomy within design activities.

This chapter's demonstrations were semi-autonomous. The *Aerial AM* system incorporated a motion capture system[25] to localise robots. Although Thread & AerialFloss robots localised using SLAM with onboard sensors, they were connected wirelessly to a central computer that ran a design-construction algorithm and made robots aware of each other's locations. With further development these demonstrations could easily be expanded to operate fully autonomously if one so desired, forgoing any offboard activities or dependencies. Technicalities aside, projects in this chapter leveraged bottom-up autonomous decision-making where individual robots operated as agents, determining their own actions in relation to each other and other perceivable information. This enabled adaptive, flexible operation, that was sufficiently robust for a variable number of robots to undertake diverse, site-specific building operations. The approach extended beyond human cognitive capabilities, to allow designers to engage large populations of robots for design and construction activities.

In addition to working in parallel, speculative Aerial Additive Manufacturing (AAM) designs also demonstrated the potential for collective robot construction systems to engage in semi-autonomous architectural design before or during construction. In each project, the robots' collective behaviour operated creatively within a scope of work set by the designer/s. Concluding projects also leveraged autonomous behaviour to facilitate situated design engagement, where solutions were integral to site-specific spatiotemporal events. Had these aerial robots commenced flight from different positions or orientations, they would have generated different design outcomes. However, these outcomes would still have fulfilled the design intent, maintaining consistency in spatial performance and aesthetic character.

In these projects, autonomous programming extended the scope of design into dynamic, future events, where the same program might play out differently in diverse environments or situations. In real terms, a designer can engage with the physical world up to hundreds of times each second (currently 50-500 Hz depending on the robot – significantly faster than human visual perception) through the orchestration of events in a collective robot system where each robot carries out behaviours authored by the designer/s. This has profound potential, extending design agency into robot populations, specific environments and sites, and dynamic events. The spatiotemporal specificity this makes possible offers a fundamentally different approach to architectural design and construction that has the potential to be sensitive to unique or varied conditions while supporting a programmed design intent. This was already alluded to in the diverse geographic and climatic regions speculative projects explored, some also strategically leveraging locally sourced materials.

▼

9.40 The Thread
Collective aerial construction: a) final installation, b) multi-agent material-physics simulation overlaid over a photo taken mid-way through aerial manufacture of installation.

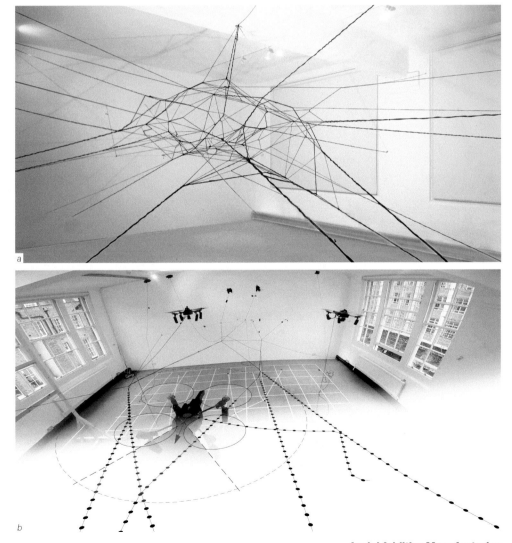

a

b

Aerial Additive Manufacturing

▶

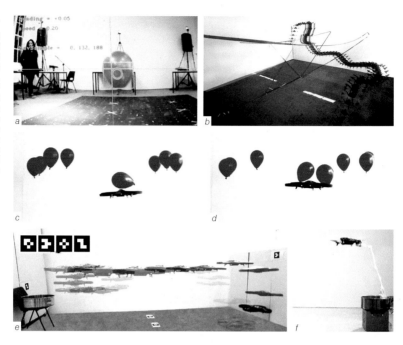

Design agency was extended by engaging other entities that have the capacity to change the physical world around us. This is an intimidating prospect, one that must carefully steer clear of ethical concerns such as James Bridle's alarm for 'Computational Thinking' where society leans too heavily on computing to solve every problem or uncritically adopts the logic of life-changing decision-making algorithms (Bridle 2019). Providing CRC is developed in an ethically risk-adverse manner, there is much to be gained from employing robot systems to perform hazardous work and to build beyond our means in hard-to-access or urgent humanitarian conditions, or in environments hostile to human physiology such as on the Moon or Mars. Distributed, parallel production can also increase productivity. In the future, similar methods might be capable of addressing unmet housing demand or expected built environment growth. Unlike traditional building approaches, in principle, collective robotic construction is economically and systemically scalable in four dimensions. Where human design and construction activities are typically short-lived, social insects such as termites leverage their collective building capabilities to engage in continual construction; maintaining, repairing and extending their habitats over time. Architectural designs manufactured by a team of robots could also engage in continuous construction to cater for constantly evolving needs. Architecture could more readily include disassembly and re-use activities or be built during specific spatiotemporal conditions, such as when weather conditions are optimal. For example, building only in the early morning during a few months of the year might be preferable for the *Aerial AM* Polar Research Center project, due to ice's material properties, adverse weather, or human comfort.

In synthesising collective robotic construction and architectural design, there is a need to establish guidelines and performance metrics that reflect the complexity of this integrated approach. In *A Review Of Collective Robotic Construction*, we introduced a metric for evaluating the efficiency of CRC systems (Petersen et al. 2019). While this metric is useful for assessing scalability and capabilities, it is also important to consider resource efficiency when evaluating architectural designs or their construction. To capture this, in a recent paper, we proposed a metric that evaluates the floor area of an *Aerial AM* building to its material volume, construction time, robot population and robot energy consumption (Stuart-Smith et al. 2023). In an industry that is increasingly concerned about product life-cycle, this metric not only encourages reductions in time, energy, and material usage, but also emphasises the importance of evaluating the population, circulation, and sequence of robot building activities.

Beyond the fact that design activities can be embodied within the behaviour of mobile robot agents, CRC extends design considerations into construction logistics, site and event adaptation, providing opportunities for architecture to engage in dynamic short-lived events or longer duration time scales of continuous construction. As discussed in Chapter Two, the Fourth Industrial Revolution is ushering in a societal reliance on autonomous systems. While it might seem like science fiction to imagine a swarm of robots constructing buildings today, projected urban population growth necessitates scalable solutions to building productivity. Looking further ahead, as we explore building in environments more hostile to our physiology such as the Moon and Mars (or must adapt to changes in climate or environment on Earth), autonomous building systems will become increasingly important. In such scenarios where societal survival or growth is key, will architects insist that a historical approach to design be executed by novel robot technologies, mandating more work and energy, or instead, seek to develop designs that consider how to optimise the material and logistics of making buildings using such robot technologies? While CRC is but one of several approaches to robotic building manufacturing, there is great potential to engage CRC to expand design agency into strange, new spatiotemporal architectural initiatives presently unimaginable.

9.42 Aerial Floss
Physical flight experiments demonstrating small-scale filament winding of: a-b) one aerial tethered space, c-d) a double-sised spatial configuration.

252

▶

9.43 Aerial Floss
*Simulated construction
behaviours adapt
to wind turbulence
to produce stable
outcomes:
a) top view of tethers
perturbed by wind
turbulence, b)
behavioural strategies
seeking to produce
stable outcomes.*

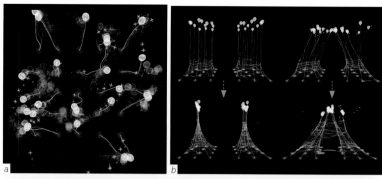

▶

9.44 Aerial Floss
*Construction is vertically
extendable by increasing
length of tethers
(simulated time L to R).*

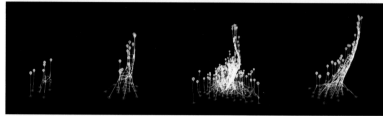

▶

9.45 Aerial Floss
*Simulated construction
of a large event-space.*

Notes

1. BIM is an abbreviation for Building Information Model, a form of information-rich 3D CAD.

2. Autonomous Manufacturing Lab, University College London, Department of Computer Science (AML-UCL). Directors: Robert Stuart-Smith and Vijay M. Pawar.

3. Autonomous Manufacturing Lab, University of Pennsylvania, Department of Architecture (AML-Penn). Director: Robert Stuart-Smith.

4. Architectural Association School of Architecture Design Research Laboratory (AA.DRL) Studio Stuart-Smith 2012-2017, Supervisor: Robert Stuart-Smith. TAs: Tyson Hosmer, Melhem Sfeir, Manos Matsis

5. Aerial AM Workshops 1 & 2. A list of workshop participants is provided in the chapter's project credits.

6. This collaborative venture was NASA-esque in its multidisciplinary ambitions and included the following scientific fields: aerial robotics, mechanical engineering, robot control, computer science, material science and structural engineering.

7. 0.4m - 1.2m diameter

8. *200g - 5kg payload*

9. *Vicon™ and Optitrack™ were utilised.*

10. *Reducing the Buildrone's velocity enabled more material to be deposited to fill in areas with insufficient build-up, while trajectory waypoint adjustments corrected for horizontal deviation in the x and y axis.*

11. *A Pixhawk PX4™ controller was utilised in lieu of the custom-developed MPC.*

12. *Deviation was less than the Buildrone delta-arm manipulator can compensate for.*

13. *Structural and geometric analysis is undertaken to identify regions with high structural stress, deflection or low surface inclination angles.*

14. *Ander et al. primarily consider self-weight and buckling, without wind-loading, but do mention the additional weight that floors, cladding etc would have on the calculations (Ander et al. n.d.).*

15. *These methods are typically used for cast-in-place concrete. Appropriated for AM, machinery might have one floor level of working height and be attached to previously manufactured layers to enable either continuous ascension or frequent detachment and re-attachment to jump up to the next floor level.*

16. *Aerial AM Polar Research Center project. See this chapter's project credits for further details.*

17. *Aerial AM Desert Shelter project. See this chapter's project credits for further details.*

18. *Aerial AM Cave Path Network project. See this chapter's project credits for further details.*

19. *Aerial AM Mountain Retreat project. See this chapter's project credits for further details.*

20. *Aerial AM Bridge project. See this chapter's project credits for further details.*

21. *Ultra-violet light*

22. *The Thread project. See this chapter's project credits for further details.*

23. *Scaling up the thickness and length of threads would increase the weight and tension in the threads, requiring larger quadcopters with bigger payloads.*

24. *Aerial Floss project. See this chapter's project credits for further details.*

25. *Vicon and Optitrack cameras were used for aspects of the research to obtain a 3D pose (position and orientation) of each aerial robot.*

▼

9.46 Aerial Floss
Single time-frame snapshot of Burning Man Festival design proposal, produced by multi-agent construction simulation.

Project Credits:

Aerial AM Shell (2022): *Robert Stuart-Smith, Patrick Danahy, Mirko Kovac, Vijay M. Pawar.*

Aerial AM Tower (2018-22): *Robert Stuart-Smith, Chris Williams, Andrew Homick, Patrick Danahy, Paul Shepherd, Vijay M. Pawar, Mirko Kovac.*

Aerial Additive Manufacturing (Aerial AM) (2016-22): *Project leaders: Mirko Kovac, Robert Stuart-Smith, Stefan Leutenegger, Vijay M. Pawar, Chris Williams, Paul Shepherd, Richard Ball. For full credits, see Zhang et al, Nature, 609 (7928), 709–717.*

Collective Aerial AM software framework (2012-23): *Autonomous Manufacturing Lab UCL (AML-UCL) Project leaders: Robert Stuart-Smith, Vijay M. Pawar. Project team: Durgesh Darekar, Sebastian Kay, Steven Hirschmann.*

Aerial AM Workshops 1 & 2 (2015): *AA.DRL and Imperial College London. Project Leaders: Robert Stuart-Smith and Mirko Kovac. Participants: Andrea Fiorentino, Deniz Topcuoglu, Eu Chian, JingJing shao, Jurij Licen, Ketao Zhang, Maria Touloupou, Nessma Al Ghoussein, Omar Ibraz, Pisak Chermprayong, Qi Cao, Sina Sareh, Sujitha Sundraraj, Talib Alhinai, Wenqian Huang, Yuang Feng.*

Aerial AM Polar Research Center (2013-14): *Architectural Association School of Architecture Design Research Laboratory (AA.DRL) Studio Stuart-Smith, Students: Quadrants – Doguscan Aladag, Tahel Shaar, Wei-Chen Ye, Juan Montiel. Supervisor: Robert Stuart-Smith. TAs: Tyson Hosmer, Manos Matsis. Advisory Consultants: AKT2 Structural Engineering.*

Aerial AM Cave Network (2016-17): *AA.DRL Studio Stuart-Smith, Students: ViManOs – Maria Touloupou, Omar Ibraz, Vikram Gaikwad. Supervisor: Robert Stuart-Smith. TAs: Tyson Hosmer, Melhem Sfeir. Advisory Consultants: AKT2 Structural Engineering.*

Aerial AM Dune Shelters (2014-15): *AA.DRL Studio Stuart-Smith, Students: Plin0os – Maria-Eleni Bali, Raíssa Carvalho Fonseca, Assad Jaffer Khan, Rithu Mathew Roy. Supervisor: Robert Stuart-Smith. TAs: Tyson Hosmer, Melhem Sfeir. Advisory Consultants: AKT2 Structural Engineering.*

Aerial AM Mountain Retreat (2015-16): *AA.DRL Studio Stuart-Smith, Students: Euphorbotics – Nessma Al Ghoussein, Qi Cao, Jurij Licen, Deniz Topcuoglu. Supervisor: Robert Stuart-Smith. TAs: Tyson Hosmer, Melhem Sfeir. Advisory Consultants: AKT2 Structural Engineering.*

Aerial AM Bridge (2013-14): *AA.DRL Studio Stuart-Smith, Students: SCL – Duo Chen, Liu Xiao, Sasila.Krishnasreni, Yiqiang Chen. Supervisor: Robert Stuart-Smith. TAs: Tyson Hosmer, Manos Matsis. Advisory Consultants: AKT2 Structural Engineering.*

The Thread (2013-14): *AA.DRL Studio Stuart-Smith, Students: Void – Alejandra Rojas, Karthikeyan Arunachalam, Maria Garcia, Melhem Sfeir. Supervisor: Robert Stuart-Smith. TAs: Tyson Hosmer, Manos Matsis. Advisory Consultants: AKT2 Structural Engineering.*

Aerial Floss (2014-15): *AA.DRL Studio Stuart-Smith, Students: AerialFloss – Kai-Jui Tsao, Patchara Ruentongdee, Yuan Liu, Qiao Zhang. Supervisor: Robert Stuart-Smith. TAs: Tyson Hosmer, Manos Matsis. Advisory Consultants: AKT2 Structural Engineering.*

Image Credits

Figures 9.1-3, 9.8-10, 9.15: *University College London, Imperial College London, University of Bath. First Published in Nature (Zhang et al. 2022).*

Figure 9.4a: *Autonomous Manufacturing Lab UCL (AML-UCL). MAP - A Mobile Agile Printer Robot for On-Site Construction (2018). Research Leads: Vijay M. Pawar and Robert Stuart-Smith. MAP Research team: Julius Sustarevas, Daniel Butters, Mohammad Hammid, George Dwyer. 2019 (Sustarevas et al. 2018).*

Figure 9.4b: *Autonomous Manufacturing Lab UCL (AML-UCL). YouWasps (2019). Thesis project: Julius Sustarevas. Supervisors: Vijay M. Pawar, Robert Stuart-Smith, David Gerber. Research team: Julius Sustarevas, K. X. Benjamin Tan (Sustarevas et al. 2019).*

Figure 9.5a,b: *©Robert Stuart-Smith & Mirko Kovac. Render by José Pareja Gómez.*

Figures 9.6-7: *University College London, Imperial College London, University of Bath. Photographs by Sarah Lever. First Published in Nature (Zhang et al. 2022).*

Figures 9.11-12: *Autonomous Manufacturing Lab (AML-UCL/AML-Penn), Robert Stuart-Smith.*

Figures 9.13-14: *University of Pennsylvania, University College London, Imperial College London, Empa (Stuart-Smith et al. 2023).*

Figures 9.16-30: *Autonomous Manufacturing Lab Penn (AML-Penn), Robert Stuart-Smith.*

Figure 9.31: Studio Robert Stuart-Smith, ViManOs student team: Maria Touloupou, Omar Ibraz, Vikram Gaikwad. Architectural Association School of Architecture Design Research Laboratory, London, 2016.

Figure 9.32:. Studio Robert Stuart-Smith, Quadrants student team: Doguscan Aladag, Juan Montiel, Tahel Shaar, Vincent Yeh. Architectural Association School of Architecture Design Research Laboratory, London, 2014.

Figure 9.33: Studio Robert Stuart-Smith, Plinθos student team: Maria-Eleni Bali, Raíssa Carvalho Fonseca, Assad Jaffer Khan, Rithu Mathew Roy. Architectural Association School of Architecture Design Research Laboratory, London, 2015.

Figure 9.34: Studio Robert Stuart-Smith, Euphorbotics student team: Nessma Al Ghoussein, Qi Cao, Jurij Licen, Deniz Topcuoglu. Architectural Association School of Architecture Design Research Laboratory, London, 2016.

Figure 9.35–36: Studio Robert Stuart-Smith, SCL student team: Liu Xiao, Sasila Krishnasreni, Duo Chen, Yiqiang Chen. Architectural Association School of Architecture Design Research Laboratory, London, 2014.

Figure 9.37–40: Studio Robert Stuart-Smith, Void student team: Karthikeyan Arunachalam, Maria García, Alejandra Rojas, Mel Sfeir. Architectural Association School of Architecture Design Research Laboratory, London, 2014.

Figures 9.41–46: Studio Robert Stuart-Smith, Aerial Floss student team: Patchara Ruentongdee, Kai-Jui Tsao and Zhang Qiao. Architectural Association School of Architecture Design Research Laboratory, London, 2015.

References

Ander, Mats, Paul Shepard, Robert Stuart-Smith, Chris Williams. "A Building of Unlimited Height." In Proceedings of the IASS Annual Symposium 2019 - Structural Membranes 2019: Form and Force, edited by Carlos Lázaro, Kai-Uwe Bletzinger, and Eugenio Oñate. Barcelona: International Association for Shell and Spatial Structures (IASS).

Andreen, David, Petra Jenning, Nils Napp, and Kirstin Petersen. 2016. "Emergent Structures Assembled by Large Swarms of Simple Robots." 36th Annual Conference of the Association for Computer Aided Design in Architecture (ACADIA).

Augugliaro, Federico, Ammar Mirjan, Fabio Gramazio, Matthias Kohler, and Raffaello D'Andrea. 2013. "Building Tensile Structures with Flying Machines." In Intelligent Robots and Systems (IROS), 2013 IEEE/RSJ International Conference On, 3487–92.

Bock, Thomas, and Thomas Linner. 2016. Site Automation. Cambridge Handbooks on Construction Robotics. Cambridge University Press.

Bonabeau, Eric, Guy Theraulaz, Jean Louis Deneubourg, Nigel R. Franks, Oliver Rafelsberger, Jean Louis Joly, and Stéphane Blanco. 1998. "A Model for the Emergence of Pillars, Walls and Royal Chambers in Termite Nests." Philosophical Transactions of the Royal Society B: Biological Sciences. doi:10.1098/rstb.1998.0310.

Bradski, Gary. 2000. "The OpenCV Library." Dr. Dobb's Journal of Software Tools.

Braithwaite, Adam, Talib Alhinai, Maximilian Haas-Heger, Edward McFarlane, and Mirko Kovač. 2018. "Tensile Web Construction and Perching with Nano Aerial Vehicles." In Robotics Research, 71–88. Springer.

Bridle, James. 2019. New Dark Age: Technology and the End of the Future. Verso Books.

Brooks, Rodney A, Pattie Maes, Maja J Matarić, and Gabriel More. 1990. "Lunar Base Construction Robots." Proceedings of the 1990 IEEE International Workshop on Intelligent Robots and Systems (IROS '90), April: 389–92. doi:10.1109/IROS.1990.262415.

Butters, Daniel, Emil T. Jonasson, Robert Stuart-Smith, and Vijay M. Pawar. 2019. "Efficient Environment Guided Approach for Exploration of Complex Environments." In IEEE International Conference on Intelligent Robots and Systems. doi:10.1109/IROS40897.2019.8968563.

Chen, An, and Mostafa Yossef. 2015. "Applicability and Limitations of 3D Printing for Civil Structures Applicability and Limitations of 3D Printing for Civil Structures." Conference on Autonomous and Robotic Construction of Infrastructure.

Chermprayong, Pisak, Ketao Zhang, Feng Xiao, and Mirko Kovac. 2019. "An Integrated Delta Manipulator for Aerial Repair: A New Aerial Robotic System." IEEE Robotics and Automation Magazine 26 (1). doi:10.1109/MRA.2018.2888911.

Cook, P, and Archigram (Group). 1999. Archigram. Princeton Architectural Press.

Cuito, Aurora, and Cristina Montes. 2003. Gaudi: Complete Works. Cologne: DuMont.
Dams, Barrie, Yuanhui Wu, Paul Shepherd, and Richard J Ball. 2017. "Aerial Additive Building Manufacturing of 3D Printed Cementitious Structures." In 37th Cement and Concrete Science Conference, London, UK United Kingdom, 11/09/17.

Dreith, Ben. 2022. "Construction Commences on BIG and ICON's Community of 3D-Printed Homes in Texas." Dezeen. November 11. https://www.dezeen.com/2022/11/11/icon-big-wolf-ranch-3d-printed-homes-austin/.

EPSRC EP/N018494/1. 2015. "EPSRC GoW: Aerial Additive Building Manufacturing: Distributed Unmanned Aerial Systems for in-Situ Manufacturing of the Built Environment." https://gow.epsrc.ukri.org/NGBOViewGrant.aspx?GrantRef=EP/N018494/1.

Farmer, Mark. 2016. "The Farmer Review of the UK Construction Model: Modernise or Die." London.

Galansino, Arturo. 2020. Tomás Saraceno: Marsilio.

Gibson, Ian, David Rosen, Brent Stucker, and Mahyar Khorasani. 2021. Additive Manufacturing Technologies. Springer International Publishing. doi:10.1007/978-3-030-56127-7.

Grassé, Pierre-Paul. 1959. "La Reconstruction Du Nid et Les Coordinations Interindividuelles Chez Bellicositermes Natalensis et Cubitermes Sp. La Théorie de La Stigmergie: Essai d'interprétation Du Comportement Des Termites Constructeurs." Insectes Sociaux 6 (1). doi:10.1007/BF02223791.

Hosmer, Tyson, Panagiotis Tigas, David Reeves, and Ziming He. 2020. "Spatial Assembly with Self-Play Reinforcement Learning." In Proceedings of the 40th Annual Conference of the Association for Computer Aided Design in Architecture: Distributed Proximities, ACADIA 2020. Vol. 1.

Hunt, Graham, Faidon Mitzalis, Talib Alhinai, Paul A Hooper, and Mirko Kovac. 2014. "3D Printing with Flying Robots." In Robotics and Automation (ICRA), 2014 IEEE International Conference On, 4493–99.

Jokic, Sasha, Petr Novikov, Shihui Jin, Stuart Maggs, Cristina Nan, and Dori Sadan. 2022. "Minibuilders." IAAC. Accessed April 14. http://robots.iaac.net/.

Kayser, Markus, Levi Cai, Sara Falcone, Christoph Bader, Nassia Inglessis, Barrak Darweesh, and Neri Oxman. 2018. "FIBERBOTS: An Autonomous Swarm-Based Robotic System for Digital Fabrication of Fiber-Based Composites." Construction Robotics 2 (1–4). doi:10.1007/s41693-018-0013-y.

Keating, Steven J., Julian C. Leland, Levi Cai, and Neri Oxman. 2017. "Toward Site-Specific and Self-Sufficient Robotic Fabrication on Architectural Scales." Science Robotics 2 (5): 15.

Klein, Georg, and David Murray. 2007. "Parallel Tracking and Mapping for Small AR Workspaces." In 2007 6th IEEE and ACM International Symposium on Mixed and Augmented Reality, ISMAR. doi:10.1109/ISMAR.2007.4538852.

Krishna, Kumar, and David A Grimaldi. 2003. "The First Cretaceous Rhinotermitidae (Isoptera): A New Species, Genus, and Subfamily in Burmese Amber." American Museum Novitates 2003 (3390). BioOne: 1–10.

Leutenegger, Stefan, Simon Lynen, Michael Bosse, Roland Siegwart, and Paul Furgale. 2014. "Keyframe-Based Visual–Inertial Odometry Using Nonlinear Optimization." The International Journal of Robotics Research 34 (3). SAGE Publications Ltd STM: 314–34. doi:10.1177/0278364914554813.

Lindsey, Quentin, Daniel Mellinger, and Vijay Kumar. 2011. "Construction of Cubic Structures with Quadrotor Teams." Robotics: Science and Systems VII. doi:10.15607/RSS.2011.VII.025.

Mataric, Maja J. 1989. "Herd Mentality." Wired 50: 46. doi:10.1136/bmj.d5820.

Mirjan, Ammar, Federico Augugliaro, Raffaello D'Andrea, Fabio Gramazio, and Matthias Kohler. 2016. "Building A Bridge with Flying Robots." In Robotic Fabrication in Architecture, Art and Design 2016, 35–47. Springer International Publishing.

Parascho, Stefana, Isla Xi Han, Samantha Walker, Alessandro Beghini, Edvard P. G. Bruun, and Sigrid Adriaenssens. 2020. "Robotic Vault: A Cooperative Robotic Assembly Method for Brick Vault Construction." Construction Robotics 4 (3–4). doi:10.1007/s41693-020-00041-w.

Petersen, Kirstin H, Nils Napp, Robert Stuart-Smith, Daniela Rus, and Mirko Kovac. 2019. "A Review of Collective Robotic Construction." Science Robotics 4 (28). Science Robotics. doi:10.1126/scirobotics.aau8479.

Petersen, Kirstin, Radhika Nagpal, and Justin Werfel. 2011. "TERMES: An Autonomous Robotic System for Three-Dimensional Collective Construction." Robotics: Science and Systems Conference VII.

Rippmann, Matthias, and Philippe Block. 2013. "Funicular Shell Design Exploration." In ACADIA 2013: Adaptive Architecture - Proceedings of the 33rd Annual Conference of the Association for Computer Aided Design in Architecture.

Sevenson, Brittney. 2015. "Shanghai-Based WinSun 3D Prints 6-Story Apartment Building and an Incredible Home." 3D Design, 3D Printing.

Staff Writer. 2013. "Kone Develops Technology for 1km-High Elevators." Construction Week. June. 17. https://www.constructionweekonline.com/business/article-22961-kone-develops-technology-for-1km-high-elevators.

Stuart-Smith, Robert. 2016. "Behavioural Production: Autonomous Swarm-Constructed Architecture." Architectural Design 86 (2). John Wiley & Sons, Ltd: 54–59. doi:10.1002/ad.2024.

Stuart-Smith, Robert, Durgesh Darekar, Patrick Danahy, Basaran Bahadir Kocer, Vijay Pawar, and Mirko Kovac. 2023. "Collective Aerial Additive Manufacturing." In Proceedings of the 42nd Annual Conference of the Association of Computer Aided Design in Architecture (ACADIA): Hybrids and Haecceities, edited by Masoud Akbarzadeh, Dorit Aviv, Hina Jamelle, and Robert Stuart-Smith, 44–55. Philadelphia,: IngramSpark.

Sustarevas, Julius, K. X. Benjamin Tan, David Gerber, Robert Stuart-Smith, and Vijay M. Pawar. 2019. "YouWasps: Towards Autonomous Multi-Robot Mobile Deposition for Construction." In IEEE International Conference on Intelligent Robots and Systems. doi:10.1109/IROS40897.2019.8967766.

Sustarevas, Julius, Daniel Butters, Mohammad Hammid, Vijay M. Pawar, George Dwyer, and Robert Stuart-Smith. 2018. "MAP - A Mobile Agile Printer Robot for on-Site Construction." In 2018 IEEE/RSJ International Conference on Intelligent Robots and Systems (IROS), 2441–48. Institute of Electrical and Electronics Engineers (IEEE). doi:10.1109/iros.2018.8593815.

Theraulaz, Guy, and Eric Bonabeau. 1995. "Coordination in Distributed Building." Science 269 (5224). The American Association for the Advancement of Science: 686.

— — —. 1999. "A Brief History of Stigmergy." Artificial Life 5 (2). MIT Press: 97–116.

Tzoumanikas, Dimos, Qingyue Yan, and Stefan Leutenegger. 2020. "Nonlinear MPC with Motor Failure Identification and Recovery for Safe and Aggressive Multicopter Flight." In 2020 IEEE International Conference on Robotics and Automation (ICRA), 8538–44. doi:10.1109/ICRA40945.2020.9196690.

Werfel, Justin, Kirstin Petersen, and Radhika Nagpal. 2011. "Distributed Multi-Robot Algorithms for the TERMES 3D Collective Construction System." In Proceedings of the IEEE/RSJ International Conference on Intelligent Robots and Systems (IROS 2011), 1–6. IEEE.

YYablonina, Maria, Marshall Prado, Ehsan Baharlou, Tobias Schwinn, and Achim Menges. 2017. "Mobile Robotic Fabrication System for Filament Structures, in Fabricate – Rethinking Design and Construction." In Fabricate Conference 2017, edited by Achim Menges, Bob Sheil, Ruairi Glynn, and Marilena Skavara, 202--209. Stuttgart: UCL Press.

Zhang, Ketao, Pisak Chermprayong, Feng Xiao, Dimos Tzoumanikas, Barrie Dams, Sebastian Kay, Basaran Bahadir Kocer, Alec Burns, Lachlan Orr, Talib Alhinai, Christopher Choi, Durgesh Dattatray Darekar, Wenbin Li, Steven Hirschmann, Valentina Soana, Shamsiah Awang Ngah, Clément Grillot, Sina Sareh, Ashutosh Choubey, Laura Margheri, Vijay M Pawar, Richard J Ball, Chris Williams, Paul Shepherd, Stefan Leutenegger, Robert Stuart-Smith & Mirko Kovac. 2022. "Aerial Additive Manufacturing with Multiple Autonomous Robots." Nature 609 (7928): 709–17. doi:10.1038/s41586-022-04988-4.

Zhang, Xu, Mingyang Li, Jian Hui Lim, Yiwei Weng, Yi Wei Daniel Tay, Hung Pham, and Quang Cuong Pham. 2018. "Large-Scale 3D Printing by a Team of Mobile Robots." Automation in Construction. doi:10.1016/j.autcon.2018.08.004.

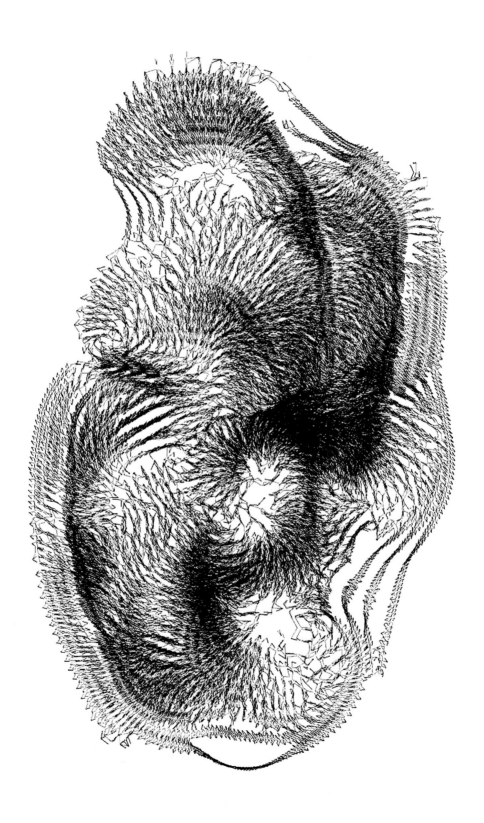

10

Behavioural Production
Semi-Autonomous Design, Manufacturing, and Construction

*B*ehavioural Production: Semi-Autonomous Approaches to Architectural Design, Robotic Fabrication and Collective Robotic Construction explored approaches to design that operated through the orchestration of spatiotemporal events within 3D modelling environments, materials, manufacturing processes, and individual and collective robotic systems. A multi-agent behavioural approach to design, manufacturing and construction was employed in diverse generative processes, authoring distinctive behavioural character. Not exactly a textbook or monograph, *Behavioural Production* is a reflection on how one can engage in design within a world increasingly dependent on autonomous systems.

Beyond the creative aspirations described in each project, more detail on the technical aspects of the work is also available in peer-reviewed scientific publications cited in each chapter, including *Nature*, *Science Robotics*, *IEEE* and others, alongside conference publications in *IROS*, *ICRA*, *ACADIA*, *CADRIA*, or *SiGRADI*. Presented design projects can also be found in journals including *AD: AD Architectural Design*, *Domus*, *Architecture Aujourd'hui*, *L'Arca*, and others. Collectively, these demonstrate a creative and scientific contribution only possible through extensive collaborative ventures, with many collaborators recognised in the acknowledgements of this book. This final chapter seeks to contextualise and map out a speculative vision for this work.

Forecasting Architecture's Trajectory

◀

10.1 bodySwarm
Single time-frame from a bodySwarm simulation. The simulation commenced with identical curve agent-bodies distributed in space before beginning to self-organise into a more complex pattern, Each agent body operates through the same behavioural rules, with the differentiated perimeter arising solely through differences in agents' local circumstances.

In 1950, the Cyberneticist Norbert Wiener predicted that with the use of a "telephone, ultrafax, or teletypewriter" an architect could effectively perform their duties remotely, designing and supervising the construction of a building located in a different country. Wiener foresaw that two-way telecommunications would alleviate the need for the architect or their drawings to be physically present on the building site (Wiener 1988, 97, location 1267). He could not have been more prescient. Seventy-two years on, following the emergence of the Internet and email, many architects work on distant, international projects. If Wiener could observe recent and emerging technological developments, what would he envisage for the capabilities and workings of an architect seventy years from today? If the last few decades of industry and academic activity demonstrate a trend, it is that enhancements in software and robot platforms will enable building design and construction to become more systemised and support more rapid and customisable production. There is also hope that this leads to higher quality and more affordable buildings with reduced embodied carbon. But what does this mean for the profession? We can surmise that the architect's scope could be potentially either more fluid or specialised.

Of note is the emergence of new building technology companies initiated by architectural practitioners. ETHZ's Gramazio Kohler Research group recently revealed a company spun out of their academic research in partnership with GoogleX, called Intrinsic (Nieva 2022). Intrinsic, is developing software to enable easier and less costly use of industrial robots. Several architects have also created specialist design and robotic fabrication companies such as Automated Architecture™ (AUAR) (Claypool et al. 2023) – a design and production startup offering robotically assembled modular architecture. Additive Manufacturing (AM) is one area where architects are currently very active. Concrete AM company XtreeE™, metal AM company MX3D™ and thermoplastic AM and design company Nagami™ were all co-founded by architects (XtreeE 2022; MX3D 2022; Jiminez-Garcia, Viguera Ochoa, and Jiménez García 2023). Although there is a risk that these companies could wind up operating solely as fabricators, they might also mimic computer company Apple™ launching their first iPhone within the telecommunications sector; potentially disrupting established industry practices. Their engagement with robotic production processes enables these companies to potentially offer more efficient and unique one-off design solutions. As exciting as this sounds, it is not so easy to achieve substantive change within the building industry. The collapse of tech start-up and modular construction company Kattera™ in 2020 illustrates the difficulties of aiming to disrupt established economic and production models. Despite attracting two billion US dollars of investment, Katerra™ found that developers and construction contractors were hesitant to break long-established ties with their own sub-contractors and material suppliers to work with an all-inclusive design and construction company (Obando 2021).

Irrespective of these initiatives' long-term success, the recent uptake of artificial intelligence (AI) software and robotic manufacturing within the building industry is likely to support a more efficient and individually customised built environment and enable alternative forms of aesthetic expression. Architecture as a discipline operates with surgical knowledge of the order and expression of space, form, geometry, tectonics, material, and ornament amongst other things. The historical canon of architecture embodies a collective body of knowledge on such topics that architects are already seeking to reinterpret through the use of AI software (Esquivel, Jaminet, and Bugni 2022; Campo, Manninger, and Carlson 2019). Such developments do not necessarily equate though to progress across all of an architect's concerns. Although an architect's capacity to address a broad range of project-specific considerations might be enhanced, there are aspects of architecture that operate outside of these. Architects through their work also creatively expand the media and impact of what is considered "architecture", as designers respond to current socio-cultural, economic and environmental needs or opportunities. Cognisant of this, as software automates and speeds up much decision-making and logistics in real estate development, design and building sectors, how will architects be sufficiently rapid and adaptive whilst not being overly reactive or repetitive when responding to accelerating market-driven pressures? How can they uphold a considered approach to design? The solution involves a partial reliance on technologies that offer more adaptation than automation affords.

The development and application of autonomous systems within diverse aspects of society today is already a key enabling technology of the Fourth Industrial Revolution. Autonomous systems can replace brute-force and generalised approaches, to support more nuanced, bespoke solutions. The architectural profession has adapted to advances from previous industrial revolutions and it stands to reason it will continue to do so again, out of necessity and opportunity. During the time this conclusion is being written, technology start-up OpenAI™ launched "ChatGPT-4", a multimodal AI software the company claims is *while less capable than humans in many real-world scenarios, exhibits human-level performance on various professional and academic benchmarks"* (OpenAI 2023). At this time when such capable AI systems are being introduced into computer and phone operating systems and apps to assist us in a broad range of tasks (Spataro 2023), one still seems to need to advocate for the benefits of leveraging degrees of autonomy within the activities of the architect or builder. This is likely to change, potentially before this passage is even published. Although ChatGPT was not utilised in the research or writing of this book, the work presented explored several different approaches in design methodology that operated vicariously through semi-autonomous systems in order to extend design capabilities and expand the agency of the architect to operate in more diverse media and spatiotemporal conditions.

The book commences with a historical and scientific context of autonomous systems. From the early automatons of Heron of Alexandria in ancient Greece (Alexandria and Woodcroft 2015), to Sheridan and Verplank's degrees of autonomy in the field of computer science (Sheridan and Verplank 1978), a case for leveraging semi-autonomous systems within design was made, advocating that such approaches can extend the designer's capabilities to operate in more responsive and nuanced ways with local site, material or temporal knowledge. Autonomous systems were not considered to be mystically intelligent (no sentience implied), but simply more adaptive than automated systems. In contrast to the unproductive and hazardous state of the construction industry presented in Chapter Two, semi-autonomous design approaches can be integrally developed in relation to manufacturing and construction logistics, increasing productivity, efficiency and design effectiveness. Such approaches align with emerging Fourth Industrial Revolution technologies including autonomous manufacturing and cyber-physical systems infrastructure that will impact the construction sector in the coming decades.

In the third chapter, through anthropologist Lambros Malafouris' concept of *'Material Agency'* (Malafouris 2008), design was recast as a generative spatiotemporal process that can be developed through engagement with a wide array of media. It was argued that each media could be considered a participant in the creative development of a design, that influences overall design cognition. Chapter Three explored this through generative approaches to material and manufacturing processes. Chapter Four introduced a multi-agent algorithmic design approach (Stuart-Smith 2011; Snooks and Stuart-Smith 2012) – which enabled degrees of design agency in 3D graphical space that was extended in Chapters Five and Six to address structural and additive manufacturing considerations.

Several creative approaches to autonomous perception were presented in Chapter Seven that leveraged computer vision and machine learning to evaluate visual character in relation to other design performance criteria (Stuart-Smith and Danahy 2022a) or to support non-expert user design intent through the employment of Deep Reinforcement Learning (Stuart-Smith and Danahy 2022b). Where Chapter Seven brought the outside world into 3D graphical space enabling situated design responses, in Chapter Eight computation was embodied within robot platforms, sensors and end-effector tools. This enabled a situated and embodied design approach that was adaptive to physical material conditions. Combined additive and subtractive manufacturing processes explored this approach (Stuart-Smith et al. 2022) in addition to work-in-progress on a multi-robot adaptive assembly software framework. While Chapter Eight showcased design operating through the act of making, Chapter Nine extended this into multi-robot construction. Semi-autonomous programming enabled design authorship to operate where the remote or pre-programmed control of mobile robot collectives was not feasible. In addition to demonstrating the first-ever untethered additive manufacturing in-flight – Aerial AM (Zhang et al. 2022), some designs also emerged from the self-organising act of collective construction (Stuart-Smith 2016).

Across most chapters in this book, custom multi-agent software was developed that operated through event-based autonomous decision-making to produce adaptive behavioural responses to several (often conflicting) design criteria. In many of the chapters, collective autonomous behaviour also gave rise to emergent design outcomes. These algorithms embedded design intent within various 3D graphics, material or robotic media, assigning these architectural agencies. Design agency thus operated within diverse domains, extending the capabilities of the designers into areas of practical and aesthetic exploration that were integral to behavioural processes of computational design, material formation, fabrication and construction. The use of degrees of autonomy in this work did not undermine or preclude regular intervention and participation by designers who engaged in intuitive, explicit design in parallel to the iterative development, testing and application of algorithmic processes. No process was considered sacred, while autonomous approaches were used to extend the designers' capabilities, not to exclude or replace them.

Behavioural Production covered many different approaches to architectural design. The work operated through the orchestration of event-based programming. This systemically open approach to design influenced the behaviour of materials, manufacturing processes, algorithms and robots, also incorporating manufacturing and construction logistics whose constraints are often difficult to reconcile within other design approaches. Where a design process is separated from other logistical considerations, it must give ground to decisions outside its remit. On the contrary, allowing design to be orchestrated within such processes opens up new creative trajectories. In the presented projects, the use of semi-autonomous programming did not undermine the designers' intent but extended it to support greater levels of engagement with the world at large.

From Generalised to Specific Design

In Chapter Two, current methods employed in the building industry were shown to be inadequate to meet present and near future production demand, nor affordability or environmental sustainability targets. The emergence of the Fourth Industrial Revolution (IR4) was discussed, and one of IR4's key driving technologies, autonomous manufacturing (Schwab 2017), was shown to have substantial potential to not only address these challenges but also to offer more expansive design and manufacturing possibilities. Where the Second Industrial Revolution (IR2) brought industrial scales of mechanical production that supported the economical mass-production of modular prefabricated building parts, the third's (IR3) development of information technologies supported the mass-customisation, and geometrical variation of building parts, and the use of 3D parametric models or BIM file-to-factory pipelines. This is most evident in building designs that exhibit degrees of visual difference in the tiling or organisation of parts across a façade or structure, or in their volumetric/formal variation. While IR3 architectural outcomes rely on narrow margins of variation within established products and systems, IR4 can support bespoke, one-off designs at industrial levels of production. This is radical. Although historically, bespoke designs were exclusively available to elite tiers of society, IR4 has the potential to enable all buildings to be realised as one-off designs without necessarily incurring time or cost penalties (see Chapter Six). Where architects operating in IR2 and IR3 paradigms had to partially generalise a design problem to suit industrial modes of production, IR4 can potentially support more distinctive and considered design responses to diverse individuals, communities, micro-climates, vegetation, species, trades, and activities, etc. To unlock such possibilities requires integrated approaches to design and IR4 autonomous manufacturing.

▶▶

10.2 Behavioural Character
Semi-autonomous processes were developed to enable design agency to operate across a wide range of media, allowing design to be expressed in spatiotemporal events. Additionally, this work aimed to create a behavioural character that embodied a qualitative gestural expression. While this character is demonstrated to achieve a degree of consistency throughout the body of work, it also reveals unique characteristics within each distinct design medium and method. This specificity in visual character operates across several scales, integrating material, formal and topological formation. Examples include: a) materialAgencies (Chapter 3), b) Helsinki Library (Chapter 3), c) Namoc (Chapter 4), d) Busan Opera (Chapter 4), e) AirBaltic Terminal (Chapter 5), f) Fibrous Tower (Chapter 5).

Situated and embodied approaches to robotics were advocated for in Chapter Eight that supported an integrated approach to design and manufacturing. Materially adaptive and bespoke production capabilities were achieved through semi-autonomous approaches to industrial robot fabrication that leveraged sensor data and computer vision feedback, together with custom approaches to robot motion and end-effector tooling. In Chapter Seven, autonomous modes of perception supported site-specific design methods. The rapidity of these processes suggested that some form of architectural agency could be explored within alternative forms of temporality. An alternative model for building construction and development was speculated on, that might operate in temporal time scales, liberating architecture from its generalist shackles by enabling it to operate more specifically for a brief moment in time. Chapter Nine not only extended temporal design considerations into multi-robot construction, but designs also leveraged autonomous approaches to multi-robot coordination, robot motion, localisation and mapping, and object recognition, and culminated in robots engaging with their physical environment to design in real-time through adaptive forms of construction. Through these developments, Chapter Nine also opened the possibility of a continuously constructed, adaptive or deconstructed architecture that employs collective robotic construction to respond to changes in people's needs or environmental conditions.

Distributed, Adaptive Manufacturing and Construction

As several decades of climate talks and economic forums have affirmed, the world is facing a global crisis of unprecedented proportions, where the Anthropocene's environmental impact and resource depletion are only to be

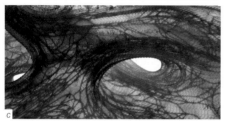

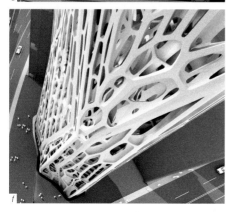

further exasperated by near-future population growth (United Nations 2022). Chapter Two detailed how current approaches to design and construction are not sufficiently productive to meet present needs. This should be alarming as we move towards a world population of 9.7 billion in 2050. An additional 2.5 billion people are set to migrate from rural areas into cities (UN DESA 2014), necessitating means to design and manufacture the built environment rapidly with greater resource and cost efficiency. To address this will take a momentous collective effort. The work presented in this book did offer a small step in this direction by presenting an approach to design, manufacturing and construction that was theoretically scalable – able to be extended to high levels of production while demonstrating unique, situated, and specific design responses that extended beyond what is achievable solely through automated technologies.

Beyond semi-autonomous approaches to computational design and robotic manufacturing, a distributed approach to off-site manufacturing and on-site construction was presented in Chapters Eight and Nine which differs from building industry approaches in use today. Both operate as a form of collective robotic construction similar to methods employed by social insects such as termites or wasps. Chapter Eight's multi-robot adaptive manufacturing framework (M-RAM) aims to enable the autonomous assembly of diverse prefabricated building modules or parts in response to a 3D Building Information Model (BIM). M-RAM aims to operate as a flexible robot work cell comprised of a variable number of fixed and/or mobile robot assets that autonomously coordinate and execute manufacturing tasks together, avoiding collisions and assembling different parts on a daily or hourly basis. In Chapter Nine, Aerial Additive Manufacturing (Aerial AM) utilised a team of aerial robots to additively manufacture in parallel in small increments over a multitude of flights. Speculative design applications of the technology were also presented, where a design emerges from the collective act of construction. While Aerial AM was developed primarily for remote, hard-to-access construction, one can speculate that Aerial AM or similar distributed robot systems could offer some applications for more general on-site construction activities also.

In both Aerial AM and M-RAM, a variable number of robots can take on one project or be distributed across several projects concurrently. This allows for on-demand, scalable building manufacturing logistics, endowed with production flexibility and dynamic resource allocation. In current construction approaches, large plant equipment (such as an AM gantry system or traditional crane), typically remains dedicated to one building site for a significant period of time. In contrast, a swarm of Aerial AM robots could converge on one building site during a stage where a project might require building across a large footprint area, and later dynamically redistribute some robots to adjacent sites when building tasks condense to a smaller footprint area that does not necessitate the same number of robots operating in parallel. Similarly, M-RAM enables potentially smaller, more flexible, distributed factories. A building project could be manufactured in one micro-factory over several months, or be farmed out to several factories distributed across a whole state or territory to be manufactured in a shorter time-frame. Robots could also be dynamically distributed within each factory to adapt to variations in task requirements across several adjacent robot work cells. A more locally distributed network of factories might also reduce building-site delivery transportation distances, reducing the carbon emissions and environmental impact of building activities. The expertise and flexibility embedded in these autonomous manufacturing and construction systems make their ubiquitous distribution and operation more feasible than automated derivations of the concept.

With an autonomous distributed manufacturing infrastructure, economies of scale would not exist to the same degree as in today's building sector. There might not be a major difference between fabricating a house through ten different manufacturing contracts versus doing it all at once. One could envisage dynamic pricing encouraging savvy builders to request manufacturing tasks to be done at hours when energy prices or material supplies are at their most economical. On-demand manufacturing and construction essentially enable design and building processes to be dynamically determined. As discussed in Chapter Six, this can also support the on-demand replacement of bespoke parts. Such principles suggest that in years to come, buildings could undergo dynamic repurposing over time or implement other interesting design and logistical considerations that offer new architectural agencies, exploring untold functions or aesthetic effects.

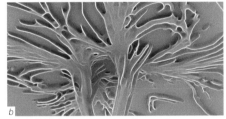

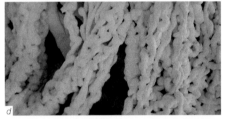

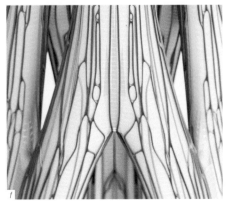

Behavioural Production

◀◀

10.3 Behavioural Character
Cont'd from Figure 10.2:
a) Tendinous bridge (Chapter 5), visual character analysis of a topoForm#2 design (Chapters 6 and 7), c) matForm ceramic prototype of topoForm#1 column (Chapter 6), d) situatedFabrications (Chapter 8), e) viscera/L (Chapter 8), f) Aerial AM Tower (Chapter 9).

An Alternative Approach to Design

At present the architectural and building industry's engagement with computational design approaches rests primarily on parametric modelling within 3D BIM software programs geared towards IR2 and IR3 manufacturing paradigms. Grounded in scientific research that extends design, manufacturing and construction capabilities, *Behavioural Production* seeks to map out a speculative path forward where designers engage with the world at large through degrees of autonomy, aligned to emerging Fourth Industrial Revolution (IR4) technologies. Rather than a one-sise-fits-all approach to design or manufacturing, *Behavioural Production* provides an opportunity for one-off engagement to take place, allowing for more dialogue not only within and between computational design, materials, manufacturing and construction systems, but also in relation to the diverse needs or desires of people, species, sites, and micro-climates. Each of these engagements does not reduce design authorship but facilitates its expansion into new territories. As outlined in the first chapter, design is not undermined by the designer's engagement with semi-autonomous processes, but extended, offering new enhanced forms of design agency, where design intent operates both directly and vicariously within spatiotemporal events.

The use of semi-autonomous processes within the book's projects also produced a behavioural character, an almost gestural qualitative expression. This design character was subject to exploration and development and was often tightly related to how each design was realised, from 3D graphics media to materials, robotic manufacturing or robot swarm construction. Chapter Four's *bodySwarm*, *NAMOC* and *Busan Opera House* projects (Figures 10.1, 10.2) assigned agency to 3D graphics assets developing related yet different character in different media (curves, surfaces, etc.). The design character of *Viscera/L* sculpted clay (Figure 10.3e) in Chapter Eight or the *Aerial AM Tower* (Figure 10.3f) in Chapter Nine, are specific not only to material or manufacturing processes but also to the spatiotemporal circumstances from which they emerged. A small change to initial conditions or events during their production would have given rise to different outcomes. A different initial shape in clay or structural input for the tower would lead to different designs. Similarly, changes to lighting conditions or clay viscosity for *Viscera/L*, or wind or specific robot flight capabilities for the tower, might impact these high-frequency, incremental design-production activities. Interestingly, due to their use of cybernetic concepts of feedback (see Chapter Eight), these design approaches afford such variable conditions and could generate successful outcomes that although different, would still embody an authored design character. In contrast to variation or difference achievable in IR2 & IR3 design approaches, *Viscera/L* or the tower's design approaches can even support chaotic divergences in design outcome. While singularities or anomalies might arise, an organisational, formal, material, or gestural character would persist. Collectively, the projects discussed in this book support one-off, bespoke design and manufacturing. They also demonstrate how machinic perception can enable materially sensitive fabrication (Chapter Eight) or how autonomous robot systems can support site-adaptive construction (Chapter Nine).

At an aesthetic or practical level, one might critique the work presented as unnecessarily complex, but in later chapters, this complexity was arguably more related to material and manufacturing processes than simpler geometrical designs sometimes are (many present-day building designs undergo geometrical rationalisation as part of a value-engineering exercise). The designs presented in this book were also frequently developed in consideration of material and structural efficiencies. Chapter Seven also sought to quantify visual character (Figure 10.3b) and relate it to material and structural metrics. It is thus not arbitrary, but a sought-after design expression that was developed in consideration of resource implications.

Through engagement with degrees of autonomous perception and action, the artefacts we design are likely to diverge from past designs as we leverage newfound freedoms, and further develop our creative agency. As Chapter Three highlighted through Lambros Malafouris' concept of *'Material Agency'* (Malafouris 2008), design cognition is responsive to the agencies of media the designer engages with. As a ceramicist's reactions to the material resistance afforded in clay inform their sculpting of a work of art, so too can the architect or designer now sense and adapt to affordances in the way they can compute, simulate and manipulate matter, or orchestrate autonomous manufacturing or construction activities. While we might not have the capacity to fully comprehend what a swarm of autonomous particles or robots might individually sense or perceive, we can develop an alternative form of design intuition within semi-autonomous processes, for how event-based rules give rise to individual and collective behaviours to determine practical and aesthetic outcomes. This, in turn, shifts our awareness of what we can achieve in design and production, but also reflects how we presently see ourselves relative to the world around us. In *What is Post-Humanism?*, Cary Wolfe describes Posthumanism as:

"a historical moment in which the decentering of the human by its imbrication in technical, medical, informatic, and economic networks is increasingly impossible to ignore, a historical development that points toward the necessity of new theoretical paradigms…" (Wolfe 2010, xv–xvi)

As Posthumanism attempts to address the decentering of the human on theoretical terms, those involved in design, manufacturing and construction need to ask similar questions (while seeking to respectively engage with diverse peoples, species, things, and environments). To venture in this direction, new forms of creative engagement are required that can draw on IR4's ability to support greater degrees of situated and embodied approaches to design, manufacture and construction through the leveraging of semi-autonomous systems. It is hoped this book encourages more activity in this space, as the benefits are still some distance away, but within our grasp.

Image Credits

Figures 10.1, 10.2c, 10.2e-f: *Kokkugia Ltd, Robert Stuart-Smith and Roland Snooks.*

Figures 10.2a: *materialAgencies: Washington University, Masters Level Fabrication Studio shared student project. Instructors: Robert Stuart-Smith & Robert Booth. Students: Guru Liu, Ruogu Liu, Lu Bai, Enrique de Solo, Shuang Jiang, Christopher Thomas Quinlan, Zhe Sun, Wing Yin (Alice) Chan, Michael Chung, Chris Moy, David J Turner, Matt White.*

Figures 10.2b, 10.3a: *Robert Stuart-Smith Design Ltd, Robert Stuart-Smith.*

Figures 10.2d: *Kokkugia Ltd, Robert Stuart-Smith.*

Figures 10.3b-c, 10.3e-f: *Autonomous Manufacturing Lab, University of Pennsylvania. Robert Stuart-Smith.*

Figures 10.3d: *SituatedFabrications Fabrication Studio. Visiting Professor: Robert Stuart-Smith. Academic Chair: Marjan Colletti. Academic Staff: Johannes Ladinig, Georg Grasser, Pedja Gavrilovic. Students: Andreas Auer, Monique banks, Marcus Bernhard, Thomas Bortondello, Marc Differding, Christophe Fanck, Konstantin Jauck, Emanuel Kravanja, Jil Medinger, Sonia Molina-Gil, Pol Olk, Michael Schwaiger, Mario Shaaya, Melina Stefanova, David Stieler, Theresa Uitz, Matthias Vinatzer, Alexandra Zeinhofer.*

References

Alexandria, Hero, and Bennet Woodcroft. 2015. Pneumatica: The Pneumatics of Hero of Alexandria. CreateSpace Independent Publishing Platform.

Campo, Matias Del, Sandra Manninger, and Alexandra Carlson. 2019. "Imaginary Plans the Potential of 2D to 2D Style Transfer in Planning Processes." In Ubiquity and Autonomy - Paper Proceedings of the 39th Annual Conference of the Association for Computer Aided Design in Architecture, ACADIA 2019.

Claypool, Mollie, Gilles Retsin, Claire-Louise McAndrew, and Manuel Jimenez-Garcia. 2023. "Automated Architecture (AUAR): Changing the Way We Build so We Can Change the Way We Live." https://automatedarchitecture.io/.

Esquivel, Gabriel, Jean Jaminet, and Shane Bugni. 2022. "The Serlio Code: Beyond Classic Language." In Projects Catalogue of the 42nd Annual Conference of the Association of Computer Aided Design in Architecture (ACADIA): Hybrids and Haecceities, 98–103. Philadelphia: Ingram Spark.

Jiminez-Garcia, Manuel, Ignacio Viguera Ochoa, and Miguel Ángel Jiménez García. 2023. "Nagami." https://nagami.design/en/.

Malafouris, Lambros. 2008. "At the Potter's Wheel: An Argument for Material Agency." In Material Agency. doi:10.1007/978-0-387-74711-8_2.

MX3D. 2022. "Joris Laarman on MX3D and Freedom of Design | MX3D." Accessed February 10. https://mx3d.com/joris-laarman-on-mx3d-and-freedom-of-design/.

Nieva, Richard. 2022. "Alphabet Launches Robotics Software Company Intrinsic from X Moonshot Lab." CNet. February 1. https://www.cnet.com/news/alphabet-launches-robotics-software-company-intrinsic-from-x-moonshot-lab/.

Obando, Sebastian. 2021. "What Does Katerra's Demise Mean for the Contech and Modular Industries?" Construction Dive. October 13. https://www.constructiondive.com/news/what-does-katerras-demise-mean-for-the-contech-and-modular-industries/608037/.

OpenAI. 2023. "GPT-4." March 14. https://openai.com/research/gpt-4.

Schwab, Klaus. 2017. The Fourth Industrial Revolution. New York: Crown Publishing Group.

Sheridan, Thomas B, and William L. Verplank. 1978. "Human and Computer Control of Undersea Teleoperators." ManMachine Systems Lab Department of Mechanical Engineering MIT Grant N0001477C0256. doi:10.1080/02724634.1993.10011505.

Snooks, Roland, and Robert Stuart-Smith. 2012. "Formation and the Rise of the Nonlinear Paradigm." Beijing, China: Tsinghua University Press.

Spataro, Jared. 2023. "Introducing Microsoft 365 Copilot – Your Copilot for Work." Official Microsoft Blog. March 16. https://blogs.microsoft.com/blog/2023/03/16/introducing-microsoft-365-copilot-your-copilot-for-work/.

Stuart-Smith, Robert. 2011. "Formation and Polyvalence: The Self-Organisation of Architectural Matter." In Ambience'11 Proceedings, edited by Annika Hellström, Hanna Landin, and Lars Hallnäs, 20–29. Boras: University of Borås.

———. 2016. "Behavioural Production: Autonomous Swarm-Constructed Architecture." Architectural Design 86 (2). John Wiley & Sons, Ltd: 54–59. doi:10.1002/ad.2024.

Stuart-Smith, Robert, and Patrick Danahy. 2022a. "Visual Character Analysis Within Algorithmic Design, Quantifying Aesthetics Relative To Structural And Geometric Design Criteria." In CAADRIA 2022, POST-CARBON - Proceedings of the 27th CAADRIA Conference - Vol. 1, Sydney, 9-15 April 2022, edited by Jeroen van Ameijde, Nicole Gardner, Kyung Hoon Hyun, Dan Luo, and Urvi Sheth, 131–40. Sydney. doi:10.52842/conf.caadria.2022.1.131.

———. 2022b. "3D Generative Design for Non-Experts: Multiview Perceptual Similarity with Agent-Based Reinforcement Learning." In Critical Appropriations - Proceedings of the XXVI Conference of the Iberoamerican Society of Digital Graphics (SIGraDi 2022), edited by PC Herrera, C Dreifuss-Serrano, P Gómez, and LF Arris-Calderon, 115–26. Lima: SIGraDi.

Stuart-Smith, Robert, Riley Studebaker, Mingyang Yuan, Nicholas Houser, and Jeffrey Liao. 2022. "Viscera/L: Speculations on an Embodied, Additive and Subtractive Manufactured Architecture." In Traits of Postdigital Neobaroque: Pre-Proceedings (PDNB), edited by Marjan Colletti and Laura Winterberg. Innsbruck: Universitat Innsbruck.

UN DESA. 2014. "World's Population Increasingly Urban with More than Half Living in Urban Areas." United Nations Department of Economic and Social Affairs. July 10. https://www.un.org/en/development/desa/news/population/world-urbanization-prospects-2014.html.

United Nations. 2022. "United Nations: Climate Action." United Nations. https://www.un.org/en/climatechange/high-level-expert-group.

Wiener, Norbert. 1988. The Human Use of Human Beings: Cybernetics And Society [Kindle IOS Version]. Hachette Books.

Wolfe, Cary. 2010. What Is Posthumanism? Posthumanities Series. University of Minnesota Press.

XtreeE. 2022. "XtreeE | The Large-Scale 3d." Https://Xtreee.Com/En/. Accessed February 10. https://xtreee.com/en/.

Zhang, Ketao, Pisak Chermprayong, Feng Xiao, Dimos Tzoumanikas, Barrie Dams, Sebastian Kay, Basaran Bahadir Kocer, Alec Burns, Lachlan Orr, Talib Alhinai, Christopher Choi, Durgesh Dattatray Darekar, Wenbin Li, Steven Hirschmann, Valentina Soana, Shamsiah Awang Ngah, Clément Grillot, Sina Sareh, Ashutosh Choubey, Laura Margheri, Vijay M Pawar, Richard J Ball, Chris Williams, Paul Shepherd, Stefan Leutenegger, Robert Stuart-Smith & Mirko Kovac. 2022. "Aerial Additive Manufacturing with Multiple Autonomous Robots." Nature 609 (7928): 709–17. doi:10.1038/s41586-022-04988-4.

Glossary

Agency:

The capacity to influence events, activities, objects, things or environments. In the context of this research, agency is encoded into software. Degrees of agency also operate in material, manufacturing and construction activities when they are employed within a generative design or production process.

Agent:

In this book, an agent typically refers to a software entity (such as a particle, curve, surface) or robot, programmed to make decisions in response to their perception of locally experienced events.

Algorithmic Design:

Design undertaken through the writing and execution of a software program. Not all algorithms are generative (see "generative" in this glossary). A design algorithm might contain a generative, parametric (associative) or procedural set of instructions.

Autonomous:

A software or robot system able to operate independently, making decisions in response to events.

Behaviour-Based/ Behavioural:

An approach to the programming of software and robot systems that builds complex adaptive behaviour through a collection of simple, bottom-up rules that are activated in relation to sensed events.

Character:

The term character is frequently used ito describe behavioural, visual or aesthetic conditions. In this context character is focused on haecceities - describing the properties or qualities of a work that make it visually distinct in comparison to other works.

Code/Coded:

A script written as an executable algorithm or software program.

Collective Behaviour:

Describes a qualitative group behaviour or pattern of behaviour that is not explicitly described in the actions or intent of the individuals. Collective behaviour is observable in virtual software agents, team-based robotics research and social species such as humans, ants, termites, etc.

Complex/Complex Systems:

Refers to a scientific notion of complexity. An object, system or pattern might exhibit complexity. Often such complexity exhibits unpredictable behaviour, and extreme sensitivity to initial conditions.

Computational Design:

Refers to design undertaken through the development of a computational method, usually involving the writing of an algorithm, but might simply involve the use of software applications to calculate various conditions.

Control (Low-Level):

A robot operates with at least one, but typically at least two levels of control. A low-level controller (program) typically instructs all hardware, determining when motors are actuated or how sensor data is processed.

Control (High-Level):

A robot might implement a high-level controller as a form of mission-planner or task manager that defines what actions the robot might take, and various autonomous decision-making principles it might follow to determine a course of action. High-level planners can address task determination, path-planning (finding a best or optimal route), exploration, team-based coordination, collective behaviour, and many other possibilities.

Drone:

A general term used to describe several types of aerial robot that are capable of flying either autonomously or by remote control. A drone is also often referred to as a "UAV".

Emergence:

Emergence describes a discernable order that arises within a system yet is not explicitly defined by the system's rules. It is an unpredictable, qualitative effect characteristically common in self-organisational systems.

Generative:

Computational, robotic or manufacturing or other processes that contribute creatively to a design outcome, beyond what is defined in their initial inputs.

Generative Design/Generative method:

See above for definition of "generative". When this term is used with no additional description it is typically referring to a generative algorithmic design process, although in the context of this book's research, a manufacturing or construction process can also be orchestrated as a generative process.

Localisation:

In the robotics community, localisation refers to the means by which a robot ascertains where it is in the world, such as its position and orientation. A robot using on-board sensors might be described as having on-board localisation or self-localisation capabilities while a robot relying on a motion capture camera system is using an external localisation method.

Multi-Agent:

See glossary definition of "Agent". Multiple agents operating as a collective, where each makes its own autonomous decisions, with its behaviour also influenced by nearby agents' behaviours.

Offline (Programming):

Describes a robot running a pre-determined program that is not altered during its execution.

Online (Programming):

Describes a robot excuting a real-time program, or being capable of live adaption in its actions.

Polyvalent:

Something is polyvalent if it incorporates multiple functions or can address multiple design criteria.

Quadcopter:

An aerial robot consisting of four propellors arranged in a square formation.

Script:

A portion of code that forms an executable algorithm or software program.

Semi-Autonomous:

Having some degree of autonomy to make some independent decisions. See "Autonomous" glossary term.

SLAM:

Simultaneous Localisation and Mapping (SLAM) is a method used in robot systems to build a map of a perceived environment while also estimating the location and orientation of the robot within that environment.

UAV:

Abbreviated term for "Unmanned Aerial Vehicle". UAV is a general term used to describe several types of aerial robot. UAVs are also often referred to as "drones".

Index

Note: Page numbers in *italics* refer to images.